CIRIL, 30 Years of Interdisciplinary Research at GANIL

Journal of Physics: Conference Series Volume 629

Caen, France
17 October 2013

ISBN: 978-1-5108-0990-1
ISSN: 1742-6588

Printed from e-media with permission by:

Curran Associates, Inc.
57 Morehouse Lane
Red Hook, NY 12571

Some format issues inherent in the e-media version may also appear in this print version.

Printed by Curran Associates, Inc. (2015)

Institute of Physics
Dirac House, Temple Back
Bristol BS1 6BE UK

Phone: 44 1 17 929 7481
Fax: 44 1 17 920 0979

techtracking@iop.org

Additional copies of this publication are available from:

Curran Associates, Inc.
57 Morehouse Lane
Red Hook, NY 12571 USA
Phone: 845-758-0400
Fax: 845-758-2634
Email: curran@proceedings.com
Web: www.proceedings.com

Table of contents

Volume 629

CIRIL, 30 years of interdisciplinary research at GANIL
17 October 2013, Caen, France

Preface

011001

CIRIL: more than 30 years of interdisciplinary research at GANIL Alicja Domaracka, Clara Grygiel, Alain Méry, Serge Bouffard and Amine Cassimi

011002

Peer review statement

Papers

012001

Primary processes: from atoms to diatomic molecules and clusters X Fléchard, L Adoui, G Ban, P Boduch, A Cassimi, J Y Chesnel, D Durand, F Frémont, S Guillous, J P Grandin, D Hennecart, E Jacquet, P Jardin, E Lamour, E Liénard, D Lelièvre, L Maunoury, A Méry, O Naviliat-Cuncic, C Prigent, J M Ramillon, J Rangama, J P Rozet, S Steydli, M Trassinelli and D Vernhet **1**

012002

Transport of fast excited ions through thin solids probed by X-ray spectroscopy E Lamour, J-P Rozet, D Vernhet, B Gervais, J-P Grandin, D Lelievre, J-M Ramillon and H Rothard **26**

012003

Ions interacting with complex molecular systems: The effect of a surrounding environment Henning Zettergren **36**

012004

Biological systems: from water radiolysis to carbon ion radiotherapy Michael Beuve, Jean-Michel Moreau, Claire Rodriguez and Etienne Testa **50**

012005

From ion-hammering to ion-shaping: an historical overview Giancarlo Rizza **61**

012006

Spectroscopic study of chemical modifications induced by swift heavy ions on polymers: the contribution of the CIRIL Platform and the CIMAP Laboratory Yvette Ngono-Ravache **82**

012007

Swift heavy ion induced electron emission from solids Hermann Rothard, Gaetano Lanzanò, Benoit Gervais, Enrico De Filippo, Michel Caron and Michael Beuve **95**

012008

Radiation effects in astrophysical ices Philippe Boduch, Emmanuel Dartois, Ana L F de Barros, Enio F da Silveira, Alicja Domaracka, Xue-Yang Lv, Maria Elisabetta Palumbo, Sergio Pilling, Hermann Rothard, Eduardo Seperuelo Duarte and Giovanni Strazzulla **102**

012009

Water radiolysis with heavy-ion beams at GANIL. Back to 20 years of investigations

G Baldacchino **112**

012010

Crystal assisted experiments for multi-disciplinary physics with heavy ion beams at GANIL Denis Dauvergne **119**

Journal of Physics: Conference Series **629** (2015) 011001 doi:10.1088/1742-6596/629/1/011001

CIRIL: more than 30 years of interdisciplinary research at GANIL

The **CIRIL: 30 years of interdisciplinary research at GANIL** meeting was held in Caen, France, in October 2013. The meeting provided an excellent opportunity to present the research performed with ion beams delivered by the GANIL facility (*Grand Accélérateur National d'Ions Lourds*, Caen, France). This proceedings volume comprises a series of reviews from different disciplines (physics, chemistry, biology, medicine, etc) of the results obtained with GANIL ion beams. Unfortunately, the issue is missing two major research domains concerning the induced damage in metals and insulators. Therefore, we refer the interested readers to the following publications: metals[1] and insulators[2].

Early non-nuclear experiments at GANIL

Over 30 years, the small pioneering team who created CIRIL (Centre for Interdisciplinary Research with Heavy Ions) has evolved into a laboratory of over 100 people, and has acquired an international reputation. The 30 years of partnership with GANIL has greatly helped the development of instrumentation dedicated to interdisciplinary research. Over this time, service to external users has remained the priority for the CIRIL–CIMAP (Centre de Recherche sur les Ions, les Matériaux et la Photonique). Technicians, engineers and researchers devote a significant part of their time to this activity and have developed instruments that are readily available to external users.

In the 1980s, two people, Jean-Pierre Briand and Yves Quéré, played an important role in the creation of a research activity in areas other than nuclear physics at GANIL: the physics of atomic collisions and the study of materials under radiation. In order to fully reach its community, they suggested that CIRIL would be independent in the research themes that it develops and supports, while remaining dependent on its GANIL collaboration to access the ion beam. This original situation has actually allowed the expansion of this research activity, although its benefits had to be underlined every time GANIL underwent a change of director. Still, we must recognize that all GANIL directors have embraced this situation, and most have understood the importance of the presence of CIRIL for the development of interdisciplinary research at GANIL. With their help, we have been able to implement investments for our communities.

[1] Dunlop A and Lesueur D 1993 Damage creation via electronic excitations in metallic targets. I. Experimental results *Rad. Eff. Def. Sol.* **126** 123

[2] Toulemonde M, Bouffard S and Studer F, Swift heavy ions in insulating and conducting oxides: tracks and physical properties 1994 *Nucl. Instrum. Methods* B **91** 1

Journal of Physics: Conference Series **629** (2015) 011001 doi:10.1088/1742-6596/629/1/011001

Figure 1: Mounting of the first atomic physics experiment on the LISE line in 1983.

One beneficial effect of CIRIL concerns the cross-fertilization of the two communities: physics of atomic collisions, and material irradiation, which previously had few opportunities to collaborate. The oral presentations of the experiment proposals at the Interdisciplinary Program Advisory Committee (iPAC) have led to better reciprocal understanding and teamwork.

In the beginning, the interdisciplinary research had access to only two equipments: IRABAT in room D5 and LISE (Ligne d'Ions Super-Épluchés) (D3–D4). IRABAT (the acronym meaning irradiation at low temperature) is a device in which the samples can be irradiated with online control of the ion flux and of its homogeneity. The material temperature can be tuned from 4 K to a few hundred °C. [3] This equipment has been thoroughly renovated and is still operational in the D1 room. It has inspired several foreign accelerator facilities' own irradiation devices. For atomic physics, LISE is an achromatic line in which highly charged ions can be produced and used for spectroscopy experiments in gas or solid phases[4]. This equipment, shared by the nuclear and atomic physicists, has been widely used for the study of collisional processes for more than fifteen years. The line was subsequently completely adapted for research on exotic ions and is now rarely used by our community.

At the beginning of CIRIL, systematic studies of defect production by GeV ions were undertaken at GANIL in a very large class of materials. These pioneering works gave many breakthrough results: for example, the effects of electronic excitation has been undoubtedly shown in metals[5] and in amorphous metallic alloys[6,7] and the evolution of the track morphology described by means of several technical measurements used concomitantly[8,9]. It is impossible to be exhaustive: in many other materials, particularly fruitful results were published for polymers[10,11], silica[12] and others.

---

[3] Jousset J C, Irradiation facilities at GANIL (Grand Accélérateur National d'Ions Lourds) 1984 *Ann. Chim. Sci. Mat.* **9** 373-376

[4] Remillieux J, Studies of non-nuclear physics at GANIL 1988 *Nucl. Instrum. Methods* B **33** 851-856

[5] Dunlop A, Lesueur D, Morillo J, Dural J, Spohr R, and Vetter J, A new damage mechanism in a crystalline metallic target irradiated with heavy gigaelectronvolts ions 1989 *C. R. Acad. Sci.* **309** 1277-1282

[6] Audouard A, Balanzat E, Fuchs G, Jousset j C, Lesueur D, and Thomé L, High-energy heavy-ion irradiations of $Fe_{85}B_{15}$ amorphous alloy: evidence for electronic energy loss effect 1987 *Europhys. Lett.* **3** 327-331

[7] Rizza; G From ion-hammering to ion-shaping: an historical overview 2015 present issue

[8] Studer F, Groult D, Nguyen N, and Toulemonde M, Threshold electronic energy loss for the creation of latent tracks in $Y_3Fe_5O_{12}$ and $BaFe_{12}O_{19}$ oxides irradiated by high energy heavy ions 1987 *Nucl. Instrum. Meth.* B **19-20** 856-859

[9] Houpert C, Studer F, Groult D, and Toulemonde M, Transition from localized defects to continuous latent tracks in magnetic insulators by high energy heavy ions: a HREM investigation 1989 *Nucl. Instrum. Meth.* B **39** 720-723

[10] Le Moël A, Duraud J P, and Balanzat E, Modifications of polyvinylidene fluoride (PVDF) under high energy heavy ion, X-ray and electron irradiation studied by X-ray photoelectron spectroscopy 1986 *Nucl. Instrum. Meth.* B **18** 59-63

[11] Ngono-Ravache Y, Spectroscopic study of chemical modifications induced by swift heavy ions on polymers: the contribution of the CIRIL Platform and the CIMAP laboratory 2015 present issue

[12] Dooryhee E, Langevin Y, Borg J, Duraud J P, and Balanzat E, Formation of paramagnetic defects in high-purity silica by high-energy ions 1988 *J. Appl. Phys.* **63** 1399-1407

At the same time, the beam-foil spectroscopies were implemented on LISE for the measurement of Lamb shift on hydrogen- or helium-like ions for QED theory testing[13] and, in the UV domain, atomic transitions were measured[14]. The experiments on charge exchange in solids led to a convincing description of the atomic processes during swift-ion flight in solids[15,16,17]. Taking advantage of the unique optical quality of the LISE line, experiments performed in channelling conditions have, for example, demonstrated the possibility of freezing the initial charge state for the well-channelled ions[18,19]. The experiments on beam-gas collision and recoil-ion spectroscopy, begun in the mid-1980s, will essentially continue with medium-energy and multi-charged low-energy ions[20,21,22].

Opportunities offered by GANIL

The years from 1983 to 1989 were years of increasing power as shown in figure 2. The number of external users gradually grew to 200 per year.

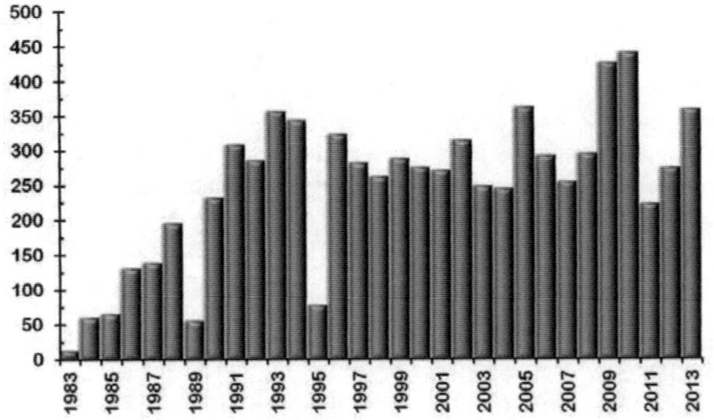

Figure 2: Number of external researchers using one of the GANIL lines since the creation of CIRIL (a user who comes twice a year is counted twice).

In 1989, major construction of the medium-energy line (SME) renewed the interdisciplinary research. Indeed, this SME line (in room D1) allows the addition of about 3000 hours of beam time for material irradiation and collision physics to the 10% of high-energy beam time. Simultaneously, IRABAT was transferred to room D1. The number of users then increased to around 300 (typically 100 different researchers coming about three times a year). Neglecting fluctuation, the annual user number has remained nearly constant for more than 20 years.

---

13 Tavernier M, Briand J P, Indelicato P, Liesen D, and Richard P, Measurement of the (1s) Lamb shift of hydrogen-like krypton 1985 *J. Phys.* B **18** L327-330

14 Martin S, Buchet J P, Buchet-Poulizac M C, Denis A, Desesquelles J, Druetta M, Grandin J P, Hennecart D, Husson X, and Lecler D, Observation of the resonant lines $2s^2S_{1/2}$-$2p^2P_{1/2,3/2}$ of lithium-like xenon 1988 *Nucl. Instrum. Meth.* B **31** 79-81

15 Rozet J P, Chetioui A, Nicolai P, Chabot M, Politis M F, Touati A, Vernhet D, Wohrer K, Stéphan C, and Grandin J P, Evolution of fine structure populations as a test of solid target effects 1991 *Z. Physik* D **21** 5331-5332

16 Lamour E, Gervais B, Rozet J P, and Vernhet D, Production and transport of long-lifetime excited states in pre-equilibrium ion-solid collisions 2006 *Phys. Rev.* A **73** 042715

17 Lamour E, Rozet J-P, Vernhet D, Gervais B, Grandin J P, Lelievre D, Ramillon J-M, Rothard H, Transport of fast excited ions through thin solids probed by X-ray spectroscopy 2015 present issue

18 Cohen C, Dural J, Gaillard M J, Genre R, Grob J J, Hage-Ali M,Kirsch R, L'Hoir A, Mory J, Poizat J C, Quéré Y, Remillieux J, Schmaus D, and Toulemonde M, High-energy heavy ions in crystals in Relativistic Channeling 1987 Plenum, New York, NY, USA, pp. 493-503

19 D. Dauvergne, Crystal assisted experiments for multi-disciplinary physics with heavy ion beams at GANIL 2015 present issue

20 Grandin J P, Hennecart D, Husson X, Lecler D, Lesteven-Vaisse I, and Lisfi D, Recoil energies of highly charged ions produced in swift ion-atom collisions as deduced from time-of-flight measurements 1988 *Europhys. Lett.* **6** 683-638

21 Fléchard X, Adoui L, Ban G, Boduch P, Cassimi A, Chesnel J-Y, Durand D, Frémont F, Guillous S, Grandin J-P, Hennecart D, Jacquet E, Jardin P, Lamour E, Liénard E, Lelièvre D, Maunoury L, Méry A, Naviliat-Cuncic O, Prigent C, Ramillon J-M, Rangama J, Rozet J-P, Steydli S, Trassinelli M and Vernhet D, Primary processes: from atoms to diatomic molecules and clusters 2015 present issue

22 Zettergren H, Ions interacting with complex molecular systems: the role of a surrounding environment 2015 present issue

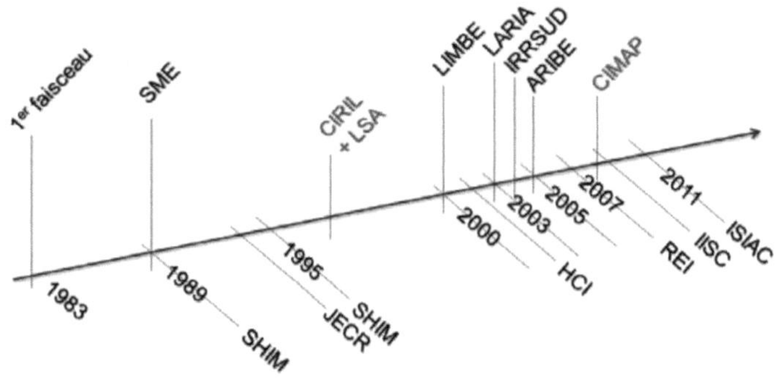

Figure 3: Main events during the lifetime of CIRIL.

Because of the exceptional opportunities offered by GANIL, CIRIL has continued to invest for the benefit of the interdisciplinary research community. First, at very low energy, with multi-charged ions delivered by ECR sources, the LIMBE facility was developed in collaboration with the Atomic Spectroscopy Laboratory, Caen, who joined CIRIL in 1999.[23]. The possibilities, initially limited to two experimental positions, were extended to five stations, with the transfer from Grenoble to Caen of the AIM facility (Multi-charged Ion Accelerator) and the hiring of two new members. The new facility, named ARIBE,[24] was inaugurated in 2005. ARIBE, associated with the European network of infrastructure ITS-LEIF, has become the mainstay of research on collisions of multiply charged ions with increasingly complex targets (molecules, C_{60}, atomic and molecular clusters, surfaces…).

IRRSUD (meaning irradiation toward south) line was built in 2004. IRRSUD takes advantage of the presence of two compact cyclotrons; however only one is used to inject towards the high-energy acceleration. The 'free' cyclotron is used for material irradiation.

Since 2005, CIRIL has therefore experiment stations in all the acceleration steps of the GANIL ions: ARIBE at the ion sources, IRRSUD after the first cyclotron (C0), SME after the second cyclotron (CSS1) and D1 at full energy. Thus, in 2013, about 6000 hours of beam time were made available to our community.

Thirty years of evolution of the interdisciplinary research

Obviously, these investments have accompanied or permitted significant changes in research topics. Spectroscopy, which was previously a major activity, disappeared in the early 1990s and was progressively replaced by research on ion--atom collision. Ten years later, with the availability of slow highly-charged ion beams, experiments on the basic processes induced by multiple electron capture on atoms, molecules and clusters have been the focus of most of the research effort. Work in this domain remains active and has led to several outcomes in radiobiology, space chemistry, ion--surface interaction and ion guiding in micro-capillaries. Figure 4 shows these thematic evolutions according to the topic of the experimental proposals.

Similarly, the research on materials under dense electronic excitation can be characterized in different phases (see figure 5). During the earlier periods, the main research effort concerned the description of the morphology of the latent track in insulators, and the study of the effect of electronic excitation on metals and metallic alloys. While the research at GANIL on metallic materials disappeared in the 2000s, research activity on latent tracks in nuclear ceramics continues. The use of ion beams as a tool to understand or control solid state properties has been the subject of many experimental proposals, up to 25% in the first half of the 1990s. Research into high-T_c

[23] Maunoury L, Leroy R, Been T, Gaubert G, Guillaume L, Lecler D, Lepoutre A, Mouton P , Pacquet J Y, Ramillon J-M, and Vicquelin R, LIMBE: A new facility for low energy beams 2002 *Rev. Sci. Instrum.* **73** 561-563

[24] Bernigaud V, Kamalou O, Ławicki A, Capron M, Maisonny R, Manil B, Maunoury L, Rangama J, Rousseau P, Chesnel J-Y, Adoui L, and Huber B A, ARIBE: a low energy ion beam facility in Caen 2008 *Publ. Astron. Obs. Belgrade* **84** 83-86

superconductors and vortex pinning by latent tracks has been particularly active during the past fifteen years, with 230 papers published in relation with an experiment performed at GANIL.

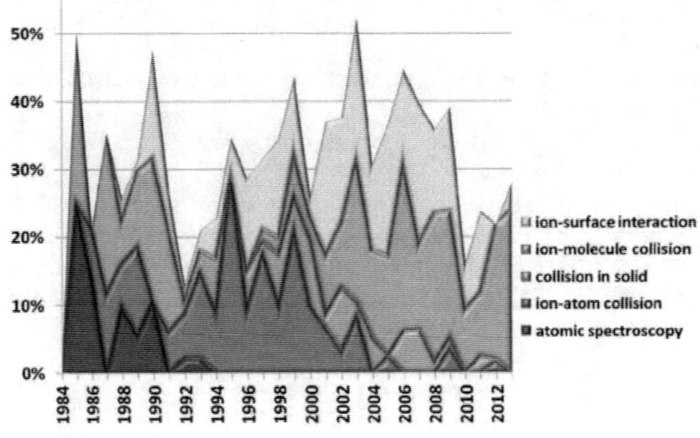

Figure 4: Evolution of the research into the physics of collisions with the ion beam of GANIL. These curves have been extracted from a compilation of all experimental proposals.

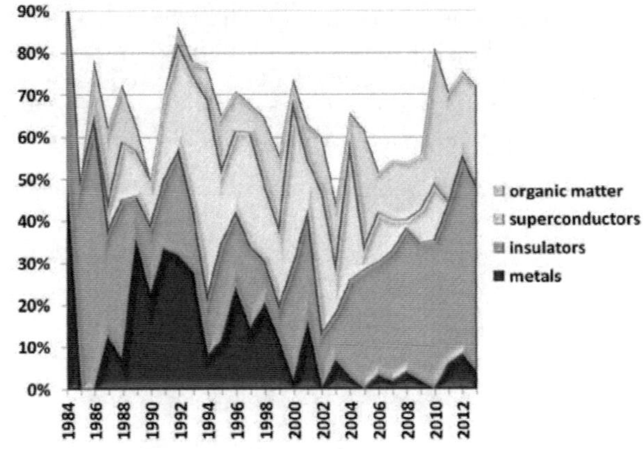

Figure 5: Evolution of the research into materials under irradiation with the ion beam of GANIL. These curves have been extracted from a compilation of all the experimental proposals.

New scientific communities have also discovered GANIL ions mostly through research where CIRIL played a major role. The organization of the JECR conference (Journées d'Etudes de la Chimie sous Rayonnement) in Caen in 1994 initiated a series of experiments on chemistry under radiation[25] with an extension to physical chemistry in the space environment[26]. The creation of a radiobiology laboratory (LARIA) jointly by CEA/DSV and CIRIL led to the growth of radiobiology activity with GANIL ions[27,28]. This laboratory is a key element of the ARCHADE project that proposes to treat cancers by hadron therapy.

In terms of users since the first beam, 70% are external, of which 70% are French. At the beginning of 2014, scientific production was substantial: more than 1400 regular papers were published and 175 PhD students have confirmed work using a GANIL beam. Most noteworthy are the 2150 authors who signed a paper linked to an experiment performed at GANIL.

These proceedings are an opportunity to acknowledge the important contribution of one of CIRIL's pioneers, Jacques Dural, who unfortunately passed away recently. This book is also dedicated to the

[25] Baldacchino G, Water radiolysis with heavy-ion beams at GANIL. Back to 20 years of investigations. 2015 present issue
[26] Boduch P, Dartois E, de Barros A L F, da Silveira E F, Domaracka A, Lv X Y, Palumbo M E, Pilling S, Rothard H, Seperuelo Duarte E and Strazzulla G, Radiation effects in astrophysical ices 2015 present issue
[27] Testard I, Radiobiology at GANIL: local project and others fields studied 2003 *Les Nouvelles du GANIL* **68** 8-13
[28] Beuve M, Moreau J-M, Rodriguez C and Testa E, Biological systems: from water radiolysis to carbon ion radiotherapy 2015 present issue

Journal of Physics: Conference Series **629** (2015) 011001 doi:10.1088/1742-6596/629/1/011001

memory of Natacha Betz, Gaetano Lanzano and Camille Cohen who have contributed through their beautiful experiments to the success of CIRIL.

Acknowledgements

The success of interdisciplinary research at GANIL is not the result of chance but owes its success to the constructive combination of several importantfactors:

- First, the unfailing dedication of the CIMAP staff for over 30 years. We gratefully thank them.
- Second, the close cooperation that CIMAP-CIRIL and GANIL have maintained. We thank GANIL for accepting an 'independent' unit inside its perimeter, and its entire staff, which makes our life easier, especially the accelerator department that provides us with high-quality beams.
- Third, thanks to all members of the different interdisciplinary program advisory committees who have worked to improve the quality of interdisciplinary research. Special thanks to the chairmen of these committees (especially J Remilleux, C Cohen, J Grihlé, G Petite, JP Nozières, D Vernhet, C Trautmann, S Klaumunzer, …).
- And of course, financial support has allowed us to invest in and maintain to the highest level our equipment. We thank CEA, CNRS, ENSICAEN, University of Caen Normandy, Regional Council of Lower Normandy, the city of Caen, ANR, Investments for the Future and Europe.

We are also grateful to all the users which, year after year, continue to submit new proposals. We would like to express our thanks to all authors of these proceedings, as well as to the reviewers for their time, effort and recommendations made during the preparation of this volume. Finally, we want to thank Jean-Claude Jousset and Jean-Pierre Grandin, two former directors, who provided the appropriate impetus to CIRIL.

Alicja Domaracka, Clara Grygiel, Alain Méry, Serge Bouffard and Amine Cassimi

Peer review statement

All papers published in this volume of *Journal of Physics: Conference Series* have been peer reviewed through processes administered by the proceedings Editors. Reviews were conducted by expert referees to the professional and scientific standards expected of a proceedings journal published by IOP Publishing.

Journal of Physics: Conference Series **629** (2015) 012001 doi:10.1088/1742-6596/629/1/012001

Primary processes: from atoms to diatomic molecules and clusters

X Fléchard[1], L Adoui[2], G Ban[1], P Boduch[2], A Cassimi[2], J Y Chesnel[2], D Durand[1], F Frémont[2], S Guillous[2], J P Grandin[2], D Hennecart[2], E Jacquet[2], P Jardin[4], E Lamour[3], E Liénard[1], D Lelièvre[2], L Maunoury[4], A Méry[2], O Naviliat-Cuncic[1], C Prigent[3], J M Ramillon[2], J Rangama[2], J P Rozet[3], S Steydli[3], M Trassinelli[3] and D Vernhet[3]

[1] LPC Caen, ENSICAEN, Université de Caen, CNRS/IN2P3, Caen, France
[2] CIMAP, CEA - CNRS - ENSICAEN, BP 5133, F-14070, Caen cedex 5, France
[3] Institut des NanoSciences de Paris, CNRS, Sorbonne Université – Pierre et Marie Curie, UMR7588, 4 Place Jussieu, 75005 Paris, France
[4] GANIL, CEA/DSM-CNRS/IN2P3, F-14070 Caen, France

E-mail: flechard@lpccaen.in2p3.fr

Abstract. This article presents a short review of the main progresses achieved at the GANIL facilities during the last thirty years in the field of ion-atom and ion-diatomic molecule collisions. Thanks to the wide range of projectile energies and species available on the different beam lines of the facility, elementary processes such as electron capture, ionization and excitation have been extensively studied. Beside primary collision mechanisms, the relaxation processes of the collision partners after the collision have been another specific source of interest. Progresses on other fundamental processes such as Young type interferences induced by ion-molecule collisions or shake off ionization resulting from nuclear beta decay are also presented.

1. Introduction

For the electronic structures of atoms and molecules, precise theoretical knowledge and high-resolution experimental data are available. But the complete understanding of dynamic processes in atomic collisions remains a challenge, due to large theoretical problems in describing time-dependent many-particle reactions, and to experimental difficulties in performing complete experiments in which all relevant quantities are accessible. Elementary collisions involving ions, atoms and molecules play an important role in many gaseous and plasma environments, where they provide both the heating and cooling mechanisms. The study of such collisions is thus not only of fundamental importance, it is also essential for the understanding of large-scale systems such as astrophysical plasmas, planetary atmospheres, gas discharge lasers, semiconductor processing plasmas, and fusion plasmas. Collisions between ions and atoms (or simple molecules) give also access to the elementary processes responsible for energy transfer in ion-matter and ion-biological molecule collisions. Complete knowledge of these elementary processes is thus of primordial importance for ion induced modification of materials as well as for radiolysis, radiotherapy and biological damages due to radiation exposure.

For inelastic collisions involving energy exchange between the collision partners, three elementary processes can be distinguished: electron capture (or charge exchange), excitation, and ionization. The relative importance of these processes depends strongly on the collision regime that is linked not only to the projectile velocity, v_p, compared to the (classical) orbital velocity of the (active) target electron, v_e, but also to the collision asymmetry, i.e. the ratio Z_p/Z_t (Z_p and Z_t are the projectile and target atomic numbers, respectively). Therefore the collision strength parameter $K = (v_e/v_p)\times(Z_t/Z_p)$ defines the collision regime such as the strong interaction regime (K>>1) where electron capture is by far the most probable process, the perturbative regime (K<<1) where ionization and excitation dominate, and the intermediate regime ($K\cong 1$) where the three processes compete, leading to many collision channels, and in which both bound and continuum states have to be taken into account for the accurate calculation of the cross sections. Of course those collision regimes can be described in terms of collision energy only, provided $Z_p \geq Z_t$. The GANIL facility of Caen, France, has the very unique feature of covering most of the energy range of interest: the ARIBE experimental hall can provide ion beams from very low (a few eV/q) to low (25 keV/q) energy, IRRSUD delivers ion projectiles in the range of 0.1 to 1 MeV/A, and by using one or two of the main cyclotrons CSS1 and CSS2, energies from 4 to 100 MeV/A are accessible. In addition, the SPIRAL1 target-ion source assembly provides low energy radioactive ions, making available the study of the electron cloud rearrangement following nuclear reactions. The GANIL is thus, for more than thirty years, a privileged experimental center to study dynamic processes with atomic and molecular targets.

The goal of the present article is to list the main advances in the field achieved thanks to the GANIL beams and to the CIMAP through the CIRIL user facility. The first section will be dedicated to experiments using atomic targets and the second section will present results obtained with small molecules and rare gas dimers. In the last section we will also discuss other fundamental dynamical processes involving ions, atoms and molecules, in which collision is the tool to reveal specific effects. The reader should be warned that this article is not an exhaustive review of ion-atom and ion-molecule collisions, what would be well beyond the scope of the present work. It is only focused on experiments performed at GANIL in the domain, with different techniques, and different specific scientific motivations that are not always naturally linked to each other.

2. Achievements in ion-atom collisions

2.1. In the strong interaction regime

Charge exchange in low-energy ion-atom collisions has been extensively studied for decades. Concurrent advances in experimental and theoretical techniques have led to a satisfying knowledge of the underlying dynamics, which is reviewed in textbooks such as [1]. However, the interest in this fundamental process has never faded out since charge exchange plays an important role in astrophysical and Tokamak plasma environments [2, 3], ion-induced radiation damage to biological cells [4, 5], and many other current applications.

2.1.1. Single electron capture.

- *Photon spectroscopy experiments*

Single electron capture leads to the population of excited states $n\ell$ of the scattered projectile ions. These excited states are below the first ionization limit of the scattered ion, thus leading to its decay by photon emission. The detection of photons emitted by the projectile and the measurement of their energy carry information on both, the spectroscopy of the resulting scattered ion, and the capture process itself.

The populated energy levels n depend mainly on the charge of the projectile and on the target ionization potential [6]. For this process, the photon spectroscopy in the range 200-600 nm is a good tool to observe Δn=1 and Δn=2 transitions [7, 8]. With the help of Hartree Fock calculations [9], it is then possible to get spectroscopic information about the energy structure of the ions produced during the collision. In the particular case of the Ar^{8+} ($1s^2 2s^2 2p^6$), a large set of targets has been used : He, H_2,

Li, Cs(6s) and Cs(6p) leading to the population of Ar VIII ($1s^2 2s^2 2p^6 n\ell$) with n=4 to 14 [8-12]. For this Ar VIII ion, many new energy levels and fine structure values have been determined [8].

Photon spectroscopy in the 200-800 nm is powerful enough to observe lines from each $n\ell$ configuration inside a same n level. The emission cross sections have been determined from the line intensities and the spectroscopic response of the optical system. The production cross sections for each $n\ell$ populated level can then be determined from the emission cross sections and the branching ratios of the emission levels in order to take into account the cascade effects on the populated levels. An example is shown in figure 1 for the Ar^{8+}-Li(2s) collision [13]. It was shown that the final $n\ell$ distributions are not statistically distributed: states with large ℓ values are strongly populated, but also states with low ℓ values (ℓ = 0, 1, 2). In most cases, the experimental production cross sections are in fair agreement with those obtained using the Classical trajectory Monte Carlo (CTMC) method (figure 1) [13-15].

At GANIL, special attention was focused on the projectile energy dependence of the production cross sections: for states with large angular momenta, no dependence has been observed, but, for a given n state (n = 8, 9 and ℓ = 0, 1, 2), the $n\ell$ cross sections increase with decreasing projectile energy. These results have then been analyzed in terms of dynamical couplings by using calculations of the electronic energies of the one-electron $(Ar^{7+}\text{-Li})^+$ system. The $n\ell$ distributions result from both the primary capture process and the post-collisional effects: these two processes depend not only on the relative importance of the radial and rotational couplings but also on the Stark mixing and on the projectile–core-electron effect. In the case of fully stripped ions colliding with lithium atoms (O^{8+}–Li), all $n\ell$ sublevels (n fixed) are affected by the Stark effect due to the Li^+ residual ion. The population of states with large ℓ values is then favored and only states with large ℓ values are populated by single-electron capture. In the case of Ar^{8+}–Li collisions, the presence of the projectile core electrons lifts the degeneracy of the $Ar^{7+}(n\ell)$ states with low ℓ values (ℓ<4) and only the states with high ℓ values are affected by the Stark effect. The $n\ell$ cross sections for ℓ<4 are strongly affected by the core-electron effect at low energy. It leads to the enhancement of the population of states with low ℓ values. The evolution of the $n\ell$ distributions with the projectile velocity depends on the evolution of the dynamical couplings: the effect of the projectile core electrons vanishes at high energies [16, 17].

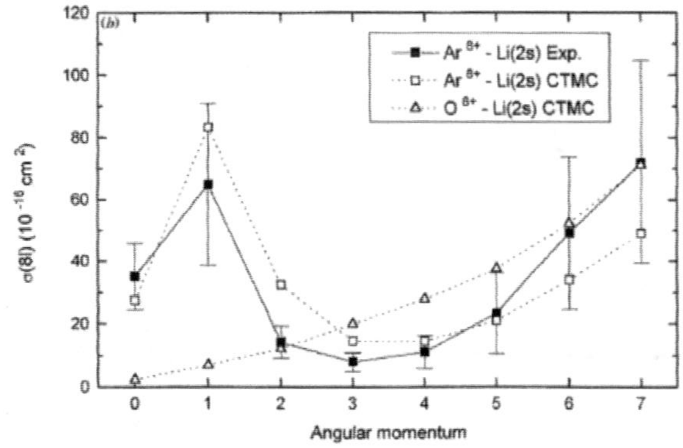

Figure 1. 8ℓ-distribution for X^{8+}-Li(2s) (X = Ar, O) collisions (projectile energy of 0.5 keV.amu^{-1}) [13].

In order to determine the m distributions, the polarization P of the emitted light has been measured. The polarization of a transition can be calculated from m capture cross sections into the different m Zeeman sublevels of the $n\ell$ upper level, taking the ion beam direction as the quantization axis. Since the spin-orbit coupling is negligible for transitions from levels such as $\ell > 2$, the formalism of Fano and Macek [18] was used. Calculated polarization ratios were obtained from these CTMC Zeeman sublevel populations. In order to make a correct comparison between the calculated and the experimental results, polarization calculations including radiative cascade effects were performed by using the method of Lin and Macek [19]. For the 8ℓ-$9\ell'$ transition for the Ar^{8+}-Li system, the high positive polarization degrees measured at the highest and lowest energies indicate that sublevels with $m = 0$ are preferentially populated. The decrease of the polarization with the collision energy indicates a broadening of the m-distributions, which is also obtained in the CTMC distributions. It is well established that the evolution of the m-distributions reflects the energy dependence of the couplings involved in the collision [20]. In the present case, the radial and rotational couplings, combined with the projectile–core electron effect and the Stark effect of the ionized target are responsible for the final $n\ell m$-distributions.

- *Kinematically complete experiments*

In the 1990s, the development of the COLTRIMS (Cold Target Recoil Ion Momentum Spectroscopy) technique [21-23], developed in parallel in Frankfurt (Germany), Manhattan (Kansas, USA) and Caen (France), provided new means to study charge exchange processes in low energy collisions. By giving access to both the Q-value and projectile scattering angle through the recoil ion momentum measurement, this technique is particularly well suited to perform kinematically complete experiments. In contrast with energy gain spectroscopy [24], the resolution in Q-value and scattering angle is not limited by the transverse emittance and by the energy dispersion of the projectile beam. In addition, the recoil ions are collected using an electric field that can be conveniently adjusted to insure 4π–detection efficiency. Combined with the high performances of the GANIL ECR source, the high detection efficiency of the COLTRIMS technique provided adequate conditions to study collisions involving fully stripped highly charged ions with charge states above 8+. The measurement of the final n-state dependence of the captured electron in coincidence with the associated scattering angle distributions could, for the first time, be performed in collision systems such as Ar^{18+} + He at 270 keV and Ne^{10+} + He at 150 keV. These challenging experiments provided new information about the reaction window and impact parameter dependence in single capture processes, and were used to test theoretical calculations based on the CTMC and the COBM (Classical Over the Barrier Model) methods. The measured n-state distributions were found to be in good agreement with both, the results of the CTMC calculations, and the $q^{0.75}$ dependence predicted by the COBM [25, 26]. Soon after, improvements of the experimental technique, such as the implementation of an inhomogeneous electric field for recoil ion collection, allowed to achieve even higher resolution on the recoil ion momentum measurement. By resolving the structures of the projectile scattering angle distributions, information about possible interferences between the different capture pathways (Stuckelberg oscillations) became available. This detailed information allowed to perform more stringent tests of close-coupling calculations using pseudo empirical exchange interaction for C^{5+} + He collisions at energies ranging from 9 keV to 90 keV [27]. In particular, the oscillations in scattering angle predicted by the calculation for 30 keV collisions were clearly observed in the experimental data. For Ne^{10+} + He collisions at 50 keV and 150 keV, the experimental scattering angle distributions obtained in the single capture process led to the validation of state-of-the-art close-coupling calculations including double capture channels and using wave functions based one-electron diatomic electron orbitals (OEDM) [28]. This constituted an important step in preparing detailed studies of the double electron capture processes. If the COLTRIMS technique proved to be a powerful tool to understand in more details the single electron capture process, it also provided new and valuable experimental results in the case of multiple electron capture, as shown in the next sections.

2.1.2. Double electron capture. When a highly charged ion collides an atomic He target at low projectile velocity ($v < 1$ a.u.), the ion may capture one or two electrons from He. Even in the case of collisions involving only two active electrons, the processes responsible for double electron capture

(DEC) were under debate during more than ten years [28-50]. Attempts to describe DEC by using independent electron models were not fully satisfactory. It has been shown that dynamic correlation effects due to the electron-electron interaction – which is not incorporated in these models – may play a decisive role during the collision [40-45]. The importance of these processes was found to depend strongly on the collision system and the projectile velocity.

DEC leads to the population of doubly excited states of the projectile ion. The production of quasi-equivalent (QE) electron configurations and non-equivalent (NE) electron configurations was investigated experimentally by means of Auger electron spectroscopy, in the case of C^{6+}, N^{7+}, O^{8+} and Ne^{10+} + He collisions [28,38-49]. At relatively high projectile velocities (~ 0.5 a.u.), QE configurations are mainly populated. This can be explained by using independent electron models. A first electron is captured and, independently of this electron, the second one is then captured. Here, electron-nucleus interaction is the main interaction responsible for such a capture. However, NE configurations are also populated, in a less extent, due to electron-electron interaction during the collision. As the projectile velocity decreases, the population of NE configurations strongly increases, showing that electron-electron interaction can play a decisive role during DEC. To identify more precisely the mechanisms responsible for the population of non-equivalent electron configurations, close-coupling calculations of DEC differential cross sections were performed and compared with experimental distributions [28, 41]. These calculations were useful to reveal the importance of the electron-electron interaction. Cold Target Recoil ion momentum spectroscopy (COLTRIMS) was also used to study the different mechanisms responsible for DEC [28, 49]. As an example, differential and doubly-differential cross sections for single-electron capture onto the levels n = 3,4,5 and 6 of $Ne^{9+}(nl)$ and double-electron capture onto series of doubly excited states 3,n and 4,n with n = 4, 5 and 6 of Ne^{8+} were obtained (figure 2) [28].

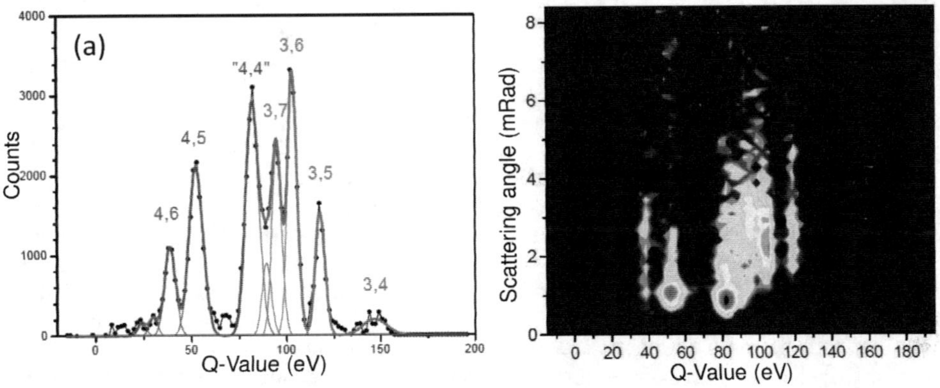

Figure 2. Cross sections differential in Q-Value (a) and doubly-differential in Q-Value and scattering angle (b) for the double electron capture process in Ne^{10+} + He collisions at 50 keV [28].

The cross sections were found to be in good agreement with state-of-the-art close-coupling calculations based on the use of OEDM orbitals. Population mechanisms for the dominantly populated double-capture channels could be deduced. In particular, it was found that even configurations of QE electrons can be populated by electron-electron interaction [28]. The population of NE electron configurations and its dependence on the collision velocity can also be attributed to post-collisional processes. These processes imply a population transfer from QE electron configurations, populated in the primary collision process, towards NE electron configurations with similar energy levels [51, 52]. The relative contribution of such processes and of the electron-electron interaction could be estimated thanks to this joint experimental/theoretical investigation. In addition, the extensive study of DEC from He made possible to derive scaling laws for electron capture in A^{q+} + He collisions at low impact velocities [45].

Finally, *K*-shell vacancy production due to dielectronic excitation has been studied in order to give direct experimental evidence for electron correlation [53,54]. The study has been performed for collisions between bare N^{7+} ions and Ne atoms. Projectile velocities ranging from 0.1 to 0.7 a.u. were investigated. The method of Auger electron spectroscopy was used to measure cross sections for producing a *K*-shell vacancy in the multi-electron target of Ne. At the low impact velocities investigated here, the cross sections were shown to be of the order of 10^{-17} cm^2. Since single excitation or ionization plays a negligible role at velocities as low as a few tenths of an atomic unit, dielectronic excitation governed by the electron-electron interaction was shown to be the unique mechanism that is responsible for the production of *K*-shell vacancies in the present collisions. The remarkable feature is that the process of dielectronic excitation is likely enough to give rise to relatively large cross sections for producing *K*-shell vacancies in very slow ion-atom collisions. Moreover, it was shown that the transfer of spectator electrons creates resonance conditions for the transfer of two active electrons via dielectronic excitation. It has been found that dielectronic excitation is particularly favored when 3 target electrons are transferred into an excited state of the projectile [54].

2.1.3. Multiple electron capture with heavy targets. Starting from the 1980s, many experiments have been performed with light ions, like C, N, Ne interacting with light targets as H_2 and He. As illustrated above, using different techniques, they lead to a quite complete understanding of the mechanisms involved in the collision [55-61]. At the same time, extended theoretical investigations have been developed. They include simple approaches such as the well-known classical over-the-barrier model (COB) [62] or the reaction window within the Landau–Zener model [63, 64], and more sophisticated coupled state calculations using basis of either atomic or molecular orbitals [65-68].

However, today, there is still a lack of data for (quasi)symmetric collisions, i.e. highly charged ions colliding on many-electron targets whilst it is indeed one of the most common fundamental processes in space such as for instance when solar wind collides with comets. Until now, the only systematic studies available are restricted to coincidence measurements between charge selected projectile and target ions [69,70] where stabilized charge exchange mechanisms are investigated.

2.1.4. Using COLTRIMS. A preliminary study of multiple capture processes in low energy (105 keV) N^{7+} + Ne collision was also attempted using the COLTRIMS technique [71]. The resolution in Q-value that can be achieved with this technique cannot compete with high resolution electron spectroscopy or photon spectroscopy, but the measurement in coincidence of the charge states of both the projectile and the recoil ion gives direct access to capture multiplicity and to the stabilization ratios of the captured electrons. In this experiment, the branching ratios for configurations populated by the single and double capture processes could be clearly resolved and quantified. For the capture of three, four and five electrons, the populated configurations could be identified but the associated branching ratios could not be accurately determined. However, it was clearly shown that triple-, quadruple-, and quintuple-electron capture populate double Rydberg states and prefer to be doubly stabilized, with two electrons remaining on the scattered projectile while the others are ejected by Auger emission. With more than two active electrons, the number of channels leading to multiple capture becomes too large to be treated theoretically within the quasimolecular description and using close coupling standard calculations. Only the COBM [63] and a semi empirical model [72] could be used to describe these processes and be compared with the experimental results. These models predict a triple-electron capture stronger than the quadruple-electron capture, what was found to be in complete disagreement with the experimental results. This implies that electon-electron interaction, not included in these oversimplified calculations, may play an important role in multiple capture.

2.1.5. What can be learned from high-resolution x-ray spectra on low energy collisions for (quasi) symmetric heavy systems? Using low-resolution x-ray detection technique, experiments have been performed [73-76], allowing to extract a parameter called the 'hardness ratio', H, (i.e. the ratio between the x-ray intensity of $n > 2 \rightarrow 1$ and $2 \rightarrow 1$ transitions) that serves as a reference to diagnose the relative abundance of constituents of intergalactic clouds and comets when interpreting astrophysical x-ray spectra [77,78]. Nevertheless, neither from ion charge state coincidence measurements nor from low-resolution x-ray spectroscopy, information on the partial $n\ell$ electron capture states is accessible. Furthermore, in the case of low-resolution x-ray measurements, as discussed recently in [79], the effect of single and multiple processes cannot be distinguished.

Of course, if a highly charged projectile collides with a many-electron target, multiple capture is possible and multiply excited projectile states can be formed. This gives rise to a complex de-excitation cascade that involves both Auger and radiative transitions. Unless one would do a "complete experiment" that measures all relevant photon and ion charge state coincidences (and possibly Auger electrons as well) it is difficult to obtain complete information on such a collision system from measurements alone. Similarly, the predictive power of theoretical calculations is limited simply because a full first-principles calculation of the many-electron dynamics is not feasible.

In a recent experiment performed at the low energy installation ARIBE at GANIL, the contribution of single-electron capture has been successfully disentangled from multiple-capture processes in the x-ray emission of an Ar projectile colliding with a N or Ar gaseous target at $v = 0.53$ a.u. [80]. By combining low-resolution spectroscopy, using Silicon Drift Detectors (SDD) but accurately calibrated in efficiency, with a complete determination of the ion beam-gas target overlap, absolute x-ray emission cross section has been extracted: a value of 11.4×10^{-15} cm^2 ±15 % has been found, in agreement with the only previous measurement performed by Tawara et *al*. [74] but with an improvement by more than a factor of 2 in the uncertainty and without referring to any external calibration.

Going a step further and taking full advantages of the high resolving power of a mosaic crystal spectrometer, additional information on single- and multi-electron processes have been achieved without applying coincidence with target charge-state detection. The whole He-like Ar Lyman series from $n = 2$ to 10 has been resolved as well as the fine structure of $n = 2 \rightarrow 1$ transitions (see figure 3) [80]. For each n value, the relative $1snp \rightarrow 1s^2$ intensity is obtained showing that the $1s2p \rightarrow 1s^2$ transition represents almost 90% of the total x-ray emission whatever the gaseous target (Ar or N). Resolving also the different $1s2\ell \rightarrow 1s$ transitions allows for a precise determination of the influence of metastable states such as the $1s2s\ ^3S_1$ state. It emphasizes that transposition of the commonly used hardness ratio, H, measured via 'laboratory ion-atom collisions' towards interpretation of spectra from comets and solar winds should be made with caution. Here, having control on the contamination from metastable Ar^{16+} ions in the incoming ion beam, an H value of 0.060 ± 0.007 has been found, approaching to the statistical limit as expected at a collision velocity of 0.53 a.u..

From the analysis of the $n = 7$–$10 \rightarrow 1$ line intensities, single-electron capture is found to occur preferentially in $n = 8$–9 with a relative n-distribution $\{Pn\}$ in agreement with the well-known Landau-Zener model [64]. In particular, $\{Pn\}$ is found to be independent on the ℓ-sublevel distribution population $\{P\ell\}$. The accurate knowledge of the Bragg-spectrometer transmission enables also to evaluate the absolute value of the single-electron capture cross section. In this case, the result is influenced by the choice of $\{P\ell\}$ and lies between σ_{single}= 4.6 10^{-15} cm^2 (flat distribution) and 12.8 10^{-15} cm^2 (statistical distribution). This result can be satisfactorily compared to the value of about 8 10^{-15} cm^2 obtained by Ali *et al* [70] from coincidence measurements of projectile–target charge exchange for Ar^{17+}on Ar at $v = 0.6$ a.u., and is, as well, in good agreement with the classical over-barrier model [63].

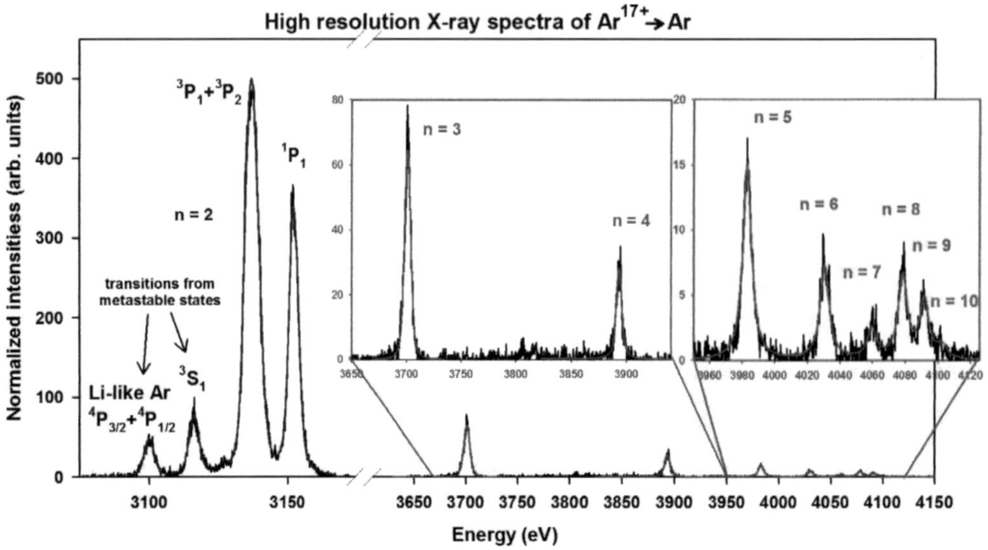

Figure 3. High-resolution spectra of Ar^{16+}x-ray transitions observed with an argon target in single collision condition. The He-like Ar 1s*n*p→ $1s^2$ with n up to 10 are visible. For $n = 2$, transitions from 1s2p $^3P_1$,$^3P_2$,$^1P_1$ to the ground level are partially resolved. In addition, M1 1s2s $^3S_1 \rightarrow 1s^{2\,1}S_0$ and Li-like Ar 1s2s2p $^4P_{1/2,3/2}$→ $1s^2$2s $^2S_{1/2}$ transitions are observed and well resolved from the 1s2p He-like transitions [80].

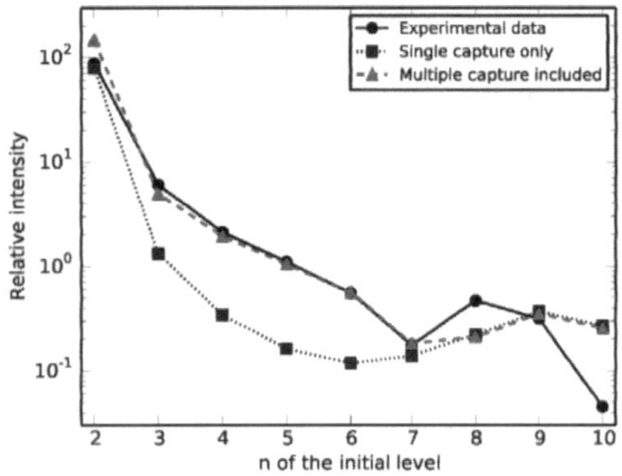

Figure 4. Experimental and calculated x-ray intensities for 1snp → $1s^2$ transitions in Ar^{16+} following capture in Ar^{17+}-Ar collisions at 0.53 a.u.

With the knowledge of $\{Pn\}$, the multi-electron capture contribution can also be estimated through the comparison between the experimental data and the expected intensities of the $n = 2 - 6 \rightarrow 1$ lines from single-electron capture using a cascade code. Those findings initiated new theoretical developments based on a two-center basis generator method (TC-BGM), in the independent electron model approximation, but including projectile states up to the 10th shell in the basis [81]. They confirm that levels of n ≥ 7 are populated only by single-capture (see figure 4), although at variance with experimental data for n=8 and 10 that is possibly due to the limited basis set used in the TG-BGM calculations. Besides, the theoretical results that include the multiple capture contributions are in very good agreement with the experimental data, while those which ignore them yield considerably lower

Journal of Physics: Conference Series **629** (2015) 012001 doi:10.1088/1742-6596/629/1/012001

intensities [82]. This demonstrates quite clearly the important role of the multiple capture processes in this type of collision systems and allows even quantifying its influence.

2.2. In the perturbative regime

For electron transitions occurring in fast ion-atom collisions, where the projectile velocity is much greater than the velocity of the active bound electron, the collision interaction is weak so that perturbative methods can be used to treat the collision dynamics [83]. The emission of slow electrons in soft collisions at large impact parameter is usually attributed to three-body effects between the projectile, the electron, and the target ion. On the other hand, for larger momentum transfer, two-body interactions involving binary-encounter processes become important. The two- and three-body processes are analogous to Compton scattering and photo-absorption, respectively. Because of their similarity to photon induced interactions, excitation and ionization by ions traveling near relativistic speeds ($v \sim 100$ a.u.) are expected to occur mainly by dipole (Δl=1) transitions. Due to the uncertainty principle $\Delta l.\Delta\theta = 1$ (θ is the emission angle relative to the beam direction), the angular momentum transfer affects the angular distribution of the emitted electrons. Low-order multipole transitions produce a broad angular distribution and vice versa.

In addition to single electron transitions, excitation and ionization can be part of a multi-electron transition. In fast ion-atom collisions, two-electron transitions can be caused by separate nucleus-electron *(n-e)* interactions, or by an *n-e* interaction followed by an electron-electron (*e-e*) interaction (e.g. reference [84] and references therein). In the former case, the process is referred to as TS2 (two-step with two projectile interactions), and in the latter case it is called TS1 (two-step with one projectile interaction). The TS1 process implies dynamic electron correlation which generally falls into two categories, corresponding to whether the first electron is emitted slowly or suddenly. For slow emission, subsequent excitation or ionization of a second electron involves the mutual scattering of two electrons, i.e., it is *dielectronic* in nature. On the other hand, sudden emission can result in a subsequent electron transition due to the change in potential seen by the second active electron as the excited system relaxes [85]. This latter type of transition is a "mean-field" effect referred to as a *shake* process.

During the past decades, ionization and excitation of helium by fast ions and photons have attracted much interest (see e.g. [86]) due to the insight it provides into dynamic correlation effects. At CIRIL, we mostly focused on the ionization and excitation of the three-electron atom of lithium. In contrast to He in which initial-state correlation is strong (both electrons initially occupy the same orbital), the Li atom has the advantage that, for double ionization, initial-state correlation is small. Furthermore, an important reason for investigating excitation in Li is the fact that both single- and double-*K*-shell vacancy production can be measured in the same experiment via Auger-electron emission (this is not possible for He).

As a first step, we investigated the cross-section ratio for double-to-single ionization for collisions of 95-MeV/u N^{7+} with He and Li target atoms [87]. The measured values for Li and He were found to be about equal (~ 0.3 %). Double ionization of He is dominated by *e-e* interactions such as shake-off and initial-state correlation. On the other hand, though double ionization of Li is dominated by *n-e* interactions (TS2), the contribution of the shake-off process is not negligible. After removal of one of the 1*s* electrons of Li, the probability for shake-off is ~ 0.38 % for the ejection of either the 1*s* or 2*s* electron [87].

Besides total cross sections, we measured doubly differential cross sections for single ionization in collisions of 95 MeV/u Ar^{18+} with atomic Li for electron emission energies ranging from 3 to 1000 eV and emission angles ranging from 25° to 155° relative to the beam direction [88-90]. The high projectile velocity provided by the GANIL facility made possible the separation of two- and three-body processes in both angular and energy distributions of the ejected electrons. Emission of the 1*s* electron was shown to be mainly influenced by three-body effects [88]. The cross section for three-body collisions rapidly decreases with the electronic energy transfer so that two-body effects dominate at high electron emission energies [88,89]. Remarkably large contributions from two-body collisions were also observed for the low-energy emission of the 2*s* electrons [89,90]. We moreover demonstrated that the node in the Li 2*s* wave function manifests itself in the angular spectra [88,90].

In parallel to ionization process studies, we devoted much attention to single and double excitation, as well as to excitation plus ionization processes. We investigated single-*K*-shell excitation and double-*K*-shell vacancy production in lithium by 95-MeV/u Ar^{18+} [91-93] and 60-MeV/u Kr^{34+} impact [94]. To separate TS1 and TS2 contributions, it was useful to vary the projectile charge-over-velocity ratio, q/v, since transition amplitudes for TS1 and shake depend linearly on q/v, whereas the transition amplitude for TS2 depends on $(q/v)^2$ [95]. Single-*K*-shell excitation was found to take place essentially by dipole transitions, leading to the excited states $1s2snp\ ^2P$. However, in the case of double *K*-shell-vacancy production, monopole transitions play a significant role, providing evidence for the *e-e* interaction. High-resolution Auger electron spectra showed that discrete doubly vacant *K*-shell states are predominantly formed from ionization plus excitation, giving rise to the $2sns\ ^{1,3}S$ and $2snp\ ^{1,3}P$ states of Li^+. The shake-up and dielectronic manifestations of the electron-electron interaction were shown to be the only mechanisms for creating the observed $2sns\ ^{1,3}S$ states. For $q/v < 2$ a.u. the formation of the $2snp\ ^{1,3}P$ states was found to be mediated only by the dielectronic process, permitting this aspect of the electron-electron interaction to be separately identified and its contribution characterized for the first time [93-96].

At high energy, for which the perturbative regime is reached, the high resolution measurements allowed by the COLTRIMS technique have led to experiments for testing the available theories on the few-body problem of the spatial and temporal evolution of mutually interacting particles. One of the most commented results has been obtained on the LISE beam line of GANIL by a collaboration between MPIK, Heidelberg and CIMAP. It was devoted to single and double ionization of He by 100 MeV/u C^{6+}. GANIL was the only place in the world to perform this experiment thanks to the available beam intensities but also to the fast bunch suppressor which could provide 1 ns ion bunches every 2 μs. Unique experimental results have been obtained concerning the double ionization [97], but surprising results on single ionization [98] have led to a 10-year-lasting discussion about the disagreement between experiment and theory (figure 5). This controversy is still open and new experiments are planned at GANIL to give a final and satisfactory answer to this disagreement.

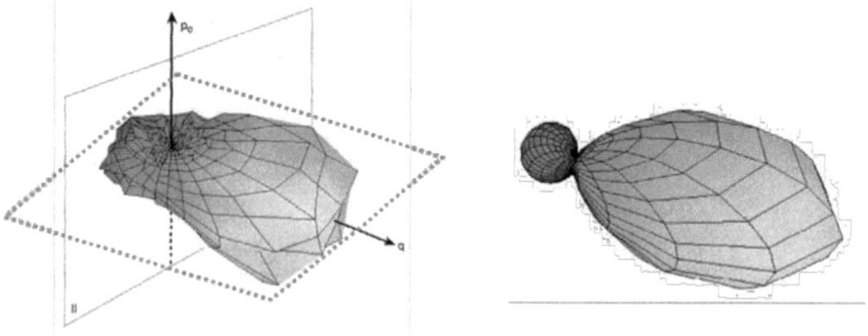

Figure 5. Electron angular distribution of helium single ionization induced by 100 MeV/u C^{6+} [99] for fixed momentum transfer q (q=0.75 u.a.) and electron kinetic energy (E=6.5 eV). Experimental results on the left, 3DW-theory on the right side.

2.3. In the intermediate regime

While significant progresses have been achieved when the perturbation induced by the target-atom is either very low or very high, as described above for instance, the full description of the dynamics of ions colliding with neutral targets in the so-named intermediate regime (i.e. K≅1, as defined in the introduction of this paper) is still lacking. Indeed, for this domain, cross sections of single-electron processes (capture, ionization or excitation) are of the same order of magnitude and interferences between all these channels cannot be neglected [100,101]. On the other hand, the role of target electrons as active partners, via capture-channel coupling and direct target electron – projectile electron interaction is not well known despite several attempts for a few specific cases (see for instance [102-105] and even recently [106,107]). The theoretical description of such many-electron systems is very complex and up to now these effects are not well taken into account even by the most sophisticated approaches. Furthermore, multiple processes, involving more than one electron of the projectile and of the target as well, start to play a role and raise much debate question even today. Experimentally, disentangling the different channels involved is also a challenge [108,109] and this collision regime remains to be explored in details whereas it corresponds to the collision domain of major interest for applications since the ion stopping power is maximum.

Performing an experiment on the medium energy line (SME; Sortie Moyenne Energie) at GANIL, with high intensity beam of Ar^{16+} at a fixed velocity, v_p= 23 a.u. (13.6 MeV/u), but using gaseous neutral targets from He to Xe, the whole range from the perturbative regime to the strong interaction regime was able to be spanned. In addition, taking full advantage of the spin selectivity in the population of the projectile excited states via the different processes involved, the single-capture (SE), filling only singlet states from the fundamental heliumlike Ar^{16+} state $^1S_0$, has been fully distinguished from the capture-ionization (CI) process, which populates triplet states as well. The same argument holds for double-excitation (DE) versus capture-excitation-ionization (CEI). Of course this type of measurements requires resolving the different final-state configurations that are separated only by a few eV. By using a high-resolution and high-transmission spectrometer, specifically designed for this experiment, cross sections of all the different processes mentioned above have been determined, including the excitation-ionization (EI) which produces hydrogenlike Ar excited states [110].

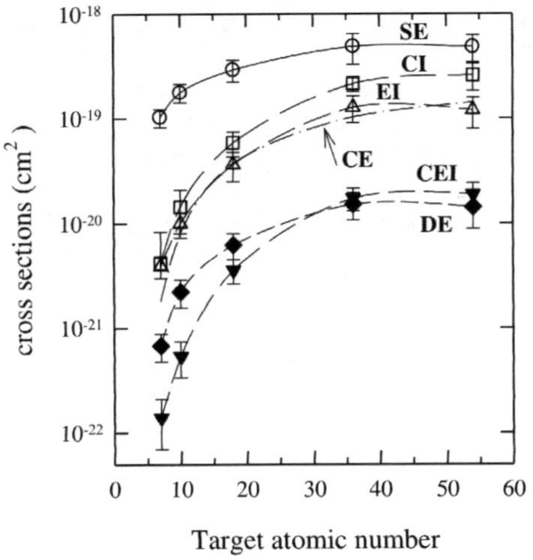

Figure 6. Single and multiple processes cross sections in n=2 for $Ar^{16+} \rightarrow Z_t$ collisions at v_p= 23 a.u. (13.6 MeV/u)

As shown in figure 6, all the processes, including both target-electrons – via capture channels – and projectile ones increase more rapidly than those involving only projectile electrons: CI for instance, which is negligible for He target (i.e. in the perturbative regime) reaches 54% of SE for Xe (i.e. when hitting the strong interaction regime). Nowadays, none of the available (even the most sophisticated) theories are able to reproduce those results on the full range of dynamics investigated here, although significant developments have been made. Only scaling laws reproducing the general relative behavior of the multiple processes compared to SE have been extracted [111, 112]. It is also worth mentioning that, even if the observed saturation of the “pure” single-excitation process, when the perturbation induced by the target-atom increases, is well predicted by the most recent models, disagreement by almost a factor of 2 was revealed starting from symmetric collisions [110]. This variance has been assigned to the role of the target-electrons, like the so-called anti-screening effect (direct projectile electron – target electron interaction), and the interferences with the capture channels which increases for high Z_t. More generally, those results demonstrated the crucial role of target

Journal of Physics: Conference Series **629** (2015) 012001 | doi:10.1088/1742-6596/629/1/012001

electrons when an ion interacts with neutral matter in the intermediate collision regime, which undeniably deserves further detailed experimental investigations.

2.3.1. Multiple ionization in the intermediate regime. The ionization process induced by swift heavy ion collisions on atomic targets is certainly the process which provided the main motivation to develop the COLTRIMS technique at beginning of the 90's [113-115]. Indeed, the ultimate goal was to get access to the scattering angle of the projectile, and hopefully to the impact parameter. As for fast collisions, the projectile scattering angle is too small to be measured directly, the idea was to infer it from the high resolution measurement of the recoiling target momentum. As mentioned above, the three pioneering groups working on this high resolution technique were the IKF, Frankfurt, Germany (H. Schmidt-Böcking), KSU, Manhattan, Kansas (C.L. Cocke) and CIMAP, Caen, France (A. Cassimi). The first results obtained with this technique have appeared in 1994 and have already evidenced the role of the electrons on the momentum balance. Their momentum has been shown to balance nearly exactly the target ion momentum making fast ion ionization resemble to photoionization.

One of the first COLTRIMS experiment performed at GANIL was aiming at the determination of the momentum vector of recoiling Ar and He target atoms ionized by Xe^{44+} (6.7 MeV/*A*) impact. From the angle of emission and recoil velocity distributions of each recoil ion charge state q (q = 1 to 7 for argon ions, q = 1 and 2 for helium ions), it was deduced that the recoil ions are mostly emitted backward with respect to the projectile beam direction (figure 7). This behaviour has been attributed to the electrons which are mainly emitted in the forward direction and thus prove, in the case of single and multiple-ionization, the importance of the role played by the electrons in the momentum balance of such a collision. Moreover, such a backward recoil ion emission proves that the interaction between the projectile and the ejected electrons on one hand and the interaction between the projectile and the target on the other hand, may not be considered as independent. As a consequence it could be stated that the two body collision model underestimates the kinetic energy deposited on the target nucleus [116].

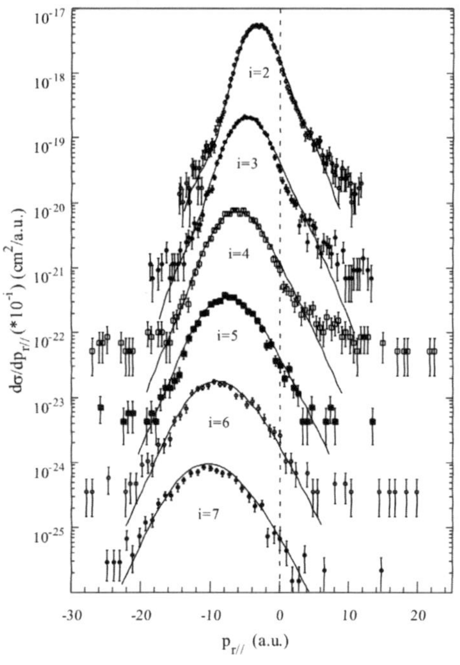

Figure 7. Experimental (dots) and calculated (lines) dσ /dp$_{r//}$ for Ar multiple ionization [116].

3. Collisions with diatomic species

3.1. Collisions with diatomic covalent molecules

3.1.1. Low energy collisions. Over the last decades, much work has been devoted to the study of slow collisions between highly charged ions and molecules [117-122]. Comparison between the collision time and the different molecular characteristic times (vibration, rotation) suggests a two-step picture for molecular fragmentation. In a first step, electron removal takes place on a fixed-in-space molecule and the transient molecular ion has the equilibrium internuclear distance of the neutral molecule. Then, in a second step, nuclear motion starts and is driven by the fragmentation dynamics. Pioneering calculations for the system $Xe^{54+} + H_2$ by means of the classical trajectory Monte Carlo method [123] suggested two main sources for the kinetic energy of each H^+ fragment: the recoil energy E_r induced by the projectile on the center of mass of the ionized target, and the fragmentation energy E_f due to a pure Coulombic dissociation. At high projectile velocities ($v_p \geq 0.5$ a.u.), the quantity E_r is negligible compared to E_f, so that the energy of the protons (~ 9.5 eV) originates only from the Coulomb explosion of the molecule at the Franck-Condon limit. In contrast, at low impact velocities ($v_p \leq 0.1$ a.u.), both slow and fast protons were predicted [123], resulting from the vector addition of the collisional momentum transfer to the center of mass of the molecule and the one due to the two-body Coulomb dissociation.

However, it has been recognized that a two-step picture is too simple [124-127]. Firstly, the fragmentation dynamics depends strongly on the primary electronic processes. In the case of multi-electron molecular targets, we have shown that electron capture can lead to the formation of inner-shell vacancies [128, 129], and thus to the excitation of the molecular target. By performing high-resolution kinetic-energy-release measurements on CO dissociation [125], the first experimental evidence that excitation of the molecular target affects the fragmentation dynamics has been provided, thus leading to significant deviations from the Coulomb explosion model. Secondly, the fragment ions may also be influenced by the outgoing projectile ion [124-127]. The influence of the projectile is similar to the well-known Post-Collision Interaction (PCI) which affects the energy spectra of Auger electrons emitted from a projectile in the field of an ionized target. The fragment energy shift due to this effect is expected to depend on the projectile charge, the projectile velocity and the orientation of the molecular target before the collision.

To get further insight into the post-collision interaction with the projectile, much effort has been devoted to investigate the fragmentation of H_2 in collisions involving O^{5+} and Xe^{23+} projectiles which strongly differ in charge, and occurring in a wide range of impact velocities (from 0.02 to 0.5 a.u.) [130, 131]. The energy distribution of the fragments was recorded at several angles with respect to the beam direction. The analysis of the spectra, combined with a classical scattering calculation, has shown that the projectile interacts strongly with each fragment. The lower the collision velocity, the lower the number of slow protons detected at forward angles (< 90°) [130, 131], since slow protons are repulsed in backward directions due to the Coulomb force induced by the projectile. Furthermore, at projectile velocities lower than 0.1 a.u., the energy distribution of the fragments exhibits two groups of peaks centered at energies which differ from the expected energy at the Franck-Condon limit ($E = 9.5$ eV): (*i*) slow fragments ($E \leq 10$ eV) are formed in soft collisions involving large impact parameters (b > 1 a.u.), while (*ii*) fast fragments originate from hard collisions between the projectile and one of the target protons [131]. This study provided the first experimental evidence that the relative contribution of hard collisions to double capture events increases when decreasing the projectile velocity [131]. We observed the same feature for electron capture following O^{5+} impact on an atomic target of helium [132], thus showing its generality.

At CIRIL, besides studies with the homonuclear H_2 target, investigations involving heteronuclear and multi-electron molecular species led to other important achievements and findings. For collisions of He^{2+} (11 keV/u) and O^{7+} (4 keV/u) on the CO molecule, the dependence of the capture process on the initial orientation of the molecule has been investigated by measuring in coincidence the full

Journal of Physics: Conference Series **629** (2015) 012001 doi:10.1088/1742-6596/629/1/012001

momentum vector of the two fragments of the dissociated molecule and the projectile charge state [122]. In that study, the KER (Kinetic Energy Release) distribution for each fragmentation channel was obtained for each outgoing projectile charge state associated not only to single capture, but also to both stabilized and autoionizing double capture processes [122]. In the case of 98-keV N^{7+} + HCl collisions, peak structures in the energy distribution of H^+ fragments ejected from HCl molecules made possible the identification of excitation and multi-electron capture channels [133]. For collisions of 30-keV N^{6+} and O^{7+} ions with water and methane molecules, the angular distribution of the fragment ions revealed nucleus-nucleus binary collision effects, with the indication of an interplay between Coulomb explosion and binary collision mechanisms [134, 135]. An unexpected anisotropy has been found for the angular distribution of H^+ emission from water [134, 135], which could be attributed to orientation sensitivity of some of the capture channels [134-137].

More recently, it has been shown that negative hydrogen ions (H^-) can be formed in molecular collisions under much more general conditions than ever expected, via a previously unrecognized process [138,139]. H^- formation has been observed in collisions between a positively-charged molecular ion (7-keV OH^+) and a neutral target (argon atom or acetone molecule). By measuring both energy and angular distributions of the emitted anions, it has been shown that H^- ions can be created via quasi-elastic two-body collisions involving a large momentum exchange between a heavy atomic center and the active H center. Hence, fragile atomic systems such as negative hydrogen ions are formed not only in soft collisions involving negligible momentum transfer, but also in hard core-core collisions via double electron grabbing by fast hydrogen fragments. For 7-keV OH^+ + Ar, a striking finding is that negative and positive hydrogen ions are emitted with very similar angular dependences [140]. Though the electron capture process is complex, the relative population of the different final charge states of the outgoing fragments can be described by simple statistical laws.

3.1.2. Medium and high energy collisions. The availability of new position sensitive detectors with the capability of multi-particle detection has opened the way to the study of the molecule fragmentation dynamics. The CIMAP has been one of the first laboratories to use this technique for the study of ion induced molecular fragmentation and took advantage of the quality GANIL ion beams [141]. As for COLTRIM spectroscopy, this imaging technique allows to measure momenta of all fragments becoming a multi-particle 4π spectrometer.

Opposite to ion-atom collisions, large momenta (up to 100 a.u.) are imparted to the molecular fragments. In this latter case, the momenta reflect the dynamics of the fragmentation process rather than the collision dynamics as it is the case for atomic targets. Indeed, they give access, with high resolution, to the Kinetic Energy Release (usually noted KER which is the sum of all fragments kinetic energy) allowing identification of the molecular excited states before the molecule dissociation. Figure 8 shows that, depending on the projectile velocity governing the dominant process (electron capture at low velocity and ionization at high velocity), the populated excited states of the CO dication are clearly different [142]. This was the first time that molecular excited states could be separated, proving the limit of the so called Coulomb explosion model which assumes that the KER is given by the Coulomb repulsion between the fragments starting at the neutral molecule atom position and holding their final charge state.

In the case of ion-molecule collisions, the target has a structure which adds new degrees of freedom such as the orientation of the molecule with regard to the beam direction. Dealing with fast projectile ions, the collision duration is much shorter (10^{-17} s) than the dissociation (10^{-14} s) and the rotation (10^{-11} s) times. This leads to two main features of this process:

- The atto-second interaction duration allows to study the free fragmentation process, without any influence of the projectile Coulomb field.
- As the rotation lasts longer than the fragmentation process, the fragment momenta keep memory of the molecule orientation at the instant of the interaction with the projectile which can thus be reconstructed.

These features have allowed to study the dependence of collision processes on the molecule orientation [141]. Strong effects have been observed and interpreted by geometric models [143, 144, 145, 146] evidencing that orientation effects become important for processes happening at impact parameters of the same order of the size of the target molecule such as inner-shell processes or multiple ionization.

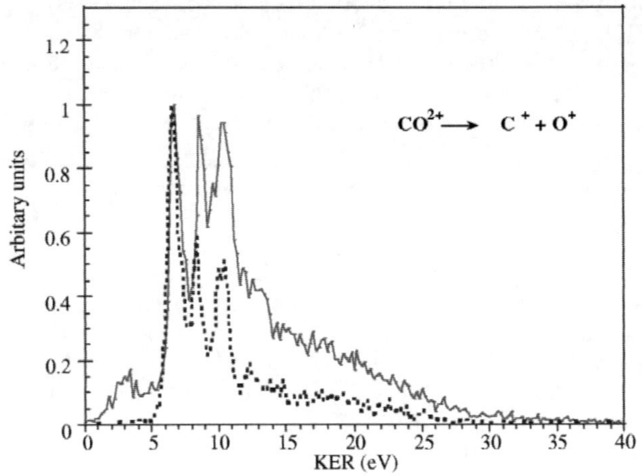

Figure 8. Comparison of the kinetic energy release distributions for the C^+ + O^+ fragmentation channel induced by 11.4 MeV u^{-1} O^{7+} (full curve) and 4 keV u^{-1} O^{7+} (dotted curve) projectiles [142].

When the target gets larger than diatomic molecules, the fragmentation dynamics starts to become more complex due to its many-body nature. Already for tri-atomic molecules such as CO_2, internal degrees of freedom increase the number of fragmentation scenarios entering into play, the fragmentation being either sequential or concerted [141]. This complicated dynamics has been accessible thanks to the latest multi-particle imaging techniques available at the end of the 90's. This dynamics can even lead to surprising isotopic effects such as those observed in the case of HDO molecules (figure 9) [147] for which the probability of losing the H^+ fragment is more than 5 times higher than for the D^+ fragment.

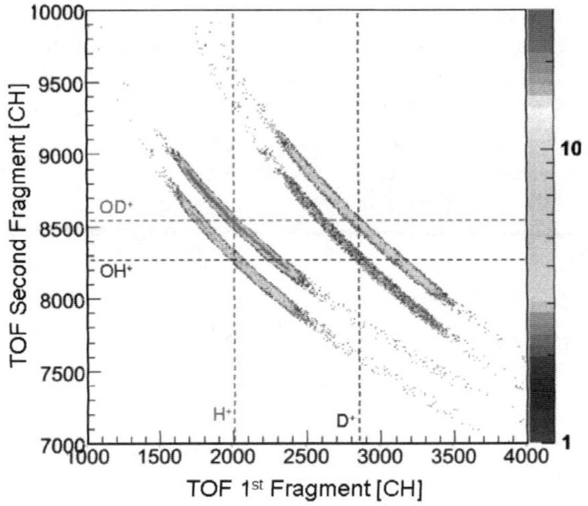

Figure 9. Two-body fragmentation channels of H_2O, D_2O and HDO mixture showing the the dominant OD^++H^+ fragmentation channel of the dication HDO^{2+} [148].

Journal of Physics: Conference Series **629** (2015) 012001 doi:10.1088/1742-6596/629/1/012001

3.2. Collisions with rare gas dimers

Very recently, rare-gas dimer targets have attracted much attention. They are characterized by a large internuclear distance and a weak molecular bond, providing a situation where the target can behave more like two neighbor independent atoms than like a covalent molecule. Photoionization experiments have already shown that new deexcitation channels open up for an ion located in the vicinity of another atom [149,150]. The new relaxation process, called inter atomic Coulombic decay (ICD) and initially predicted by Cederbaum *et al* [151], has been observed for biologically relevant systems such as water clusters [152] and a vast number of rare-gas clusters [153]. This process is thus an additional source of low-energy electrons that can play a role in radiation damage to healthy tissues.

In this context, the contribution of ICD to the production of low-energy electrons induced by ion impact had to be investigated. In ionizing ion-atom collisions at medium and high energy collisions, the dominant contribution to the electron energy distribution results from distant collisions with small energy transfers which can be approximated by an exponential decay. Recent experiments performed at GANIL with 11.37 MeV/u S^{14+} ions colliding on neon dimers have shown a large contribution of low-energy electron production unambiguously attributed to ICD [154]. Combined with complementary experiments using He^{+} and He^{2+} projectiles in the 0.1-0.2 MeV/u energy range on neon and argon dimer targets, the results suggest that ICD is omnipresent and a major contributor of low-energy electron emission in fast ion collisions with loosely bound targets [154].

For lower energy collisions between HCIs (highly charged ions) and molecular targets, multiple electron capture becomes the dominant primary process. For molecular targets with a strong covalent character, for which the valence electronic orbitals are delocalized over the whole molecule, multiple ionization leads usually to dissociation into equally charged fragments [155-157]. With rare gas dimer targets, if one site of the dimer is multiply ionized, near isolation of the two atoms may lead to weaker and slower charge rearrangement. This unique feature may allow to investigate how the electrons are captured at the instant of the collision and to study electron dynamics in charge transfer collisions with HCIs. In an experiment involving low energy Ar^{9+} projectiles colliding on Ar_2, the yields associated to the different fragmentation channels $Ar^{+} + Ar^{+}$, $Ar^{2+} + Ar^{+}$, $Ar^{3+} + Ar^{+}$ and $Ar^{2+} + Ar^{2+}$ were precisely measured. In contrast with the results previously obtained with covalent molecules, it was shown that asymmetric fragmentation channels are favored [158,159]. In addition, the kinetic energy release (KER) spectrum of the $Ar^{+} + Ar^{+}$ fragmentation channel allowed to identify one-site double capture process populating bound transient states that can only decay through radiative charge transfer (RCT). This behavior was interpreted as due to low electron-mobility within the molecular ions resulting from the collision.

This finding motivated a more detailed study of the collision system with the measurement of the correlation between the projectile scattering angle and the molecular ion orientation, as a function of the fragmentation charge sharing. As illustrated in the figures 10.a and 10.b, the projectile scattering angle ϕ_{proj} is given by the transverse momentum exchange due to the Coulomb repulsion between the collision partners. It is thus closely related to the impact parameter vector $\vec{b}$ in the molecular frame and to the final charge on the two sites of the dimer. The experimental results provided clear evidence that projectiles distinguish each atom in the target and, that electron capture from near-site atom is favored [160]. Monte Carlo calculations based on the classical over-the-barrier model, with dimer targets represented as two independent atoms, were compared to the data. These calculations, in good agreement with the experimental data, give a new insight into the dynamics of the collision by providing, for the different electron capture channels, the two-dimensional probability maps $p(\vec{b})$ [160].

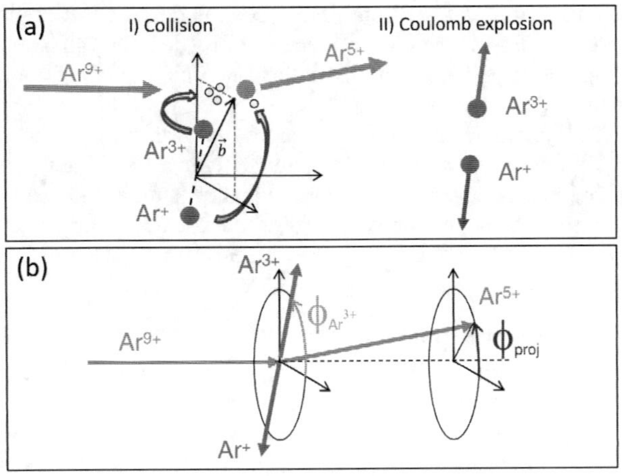

Figure 10. (Color online) Schematic view of the multiple electron capture from Ar_2 by Ar^{9+}projectiles resulting in the Ar^{3+}+ Ar^{+} asymmetric fragmentation channel (a). Representation of the scattering angle ϕ_{proj} and of the angle of emission ϕ_{Ar3+} of the most charged fragment in the plane transverse to the beam axis (b).

4. Probing other fundamental processes

4.1. Young-type interferences induced by ion-molecule collisions

Since 2001, many experiments have been devoted to the evidence for oscillations in angular distribution and energy distribution of electrons emitted in collisions of highly charged, fast and slow ions [161-178] and electrons [179] with H_2 molecules. First, structures attributed to interference effects in the electron emission spectra from H_2, which are similar to those observed in Young's two-slit experiment, were observed [161-166]. For ion-induced electron emission the interference structures were difficult to observe. Therefore theory has been used to provide guidance in the search for the interference phenomena. Fast projectiles have been found to be essential (v > 10 a.u.), since they enhance dipole-like transitions which, in turn, are responsible for the interference effects. Various features predicted by the model calculations are in accordance with the experimental results: (i) interference effects do not cancel when an averaging is performed over the orientation of the H_2 internuclear axis, (ii) interference effects are manifested by a sinusoidal-like oscillation occurring in an energy range up to 250 eV. Evidence is provided for the fact that the oscillation frequency of the interference pattern varies significantly with the electron ejection angle. The theoretical results based on Born approximation confirm this behavior. However, the theory cannot explain the high frequency observed at high emission angles [166].

A few years later, interferences caused by a single electron impacting on an independent double-center scatterer, which plays the role of an atomic-size double-slit system, were experimentally evidenced for the first time [167-178]. The electron originates from the autoionization of doubly excited *2lnl'* ($n \geq 2$) configurations of He following a double charge exchange process by He^{2+} ions impinging on H_2 molecules. Well-defined oscillations could be observed in the angular distribution of the electrons emitted towards the receding H protons. The presence of these oscillations provides the first experimental demonstration that a single electron interferes with itself [167]. This is analogous to the famous "thought" experiment imagined and discussed by Feynman in 1963, in which the quantum nature of the electron was illustrated by making it traverse an atomic-size double-slit arrangement.

An interesting aspect of electron interferences was also evidenced [174]. Instead of investigating the total intensity of undiscerned *2lnl'* ($n \geq 2$) autoionization configurations, focus has been put on a single $2s^2\,{}^1S$ line. Its maximum and linewidth have been determined at angles ranging from 120° to 160°, where interferences are expected to occur. This detailed analysis revealed well-defined oscillations in the angular dependence of both the maximum intensity and the linewidth. In the investigated angular range, the maximum oscillates in phase with the total intensity, showing that the oscillations of the total intensity are mainly due to the variations of the line maximum. More surprisingly, the $2s^2\,{}^1S$ linewidth was found to strongly oscillate in counter phase with the maximum, a fact that can be explained by means of simple theoretical arguments. These results not only provide a Feynman-type demonstration of the presence of a nanoscale Young-type interference of single electrons but also complete and reinforce the analysis performed previously on the undiscerned $2lnl'$ ($n \geq 2$) configurations.

4.2. Electron shakeoff in nuclear beta decay

Electron shakeoff (SO) is a fundamental atomic process resulting from the sudden change of the central potential in which a bound electron is ejected into the continuum. This monopole ionization may be due to the creation of a vacancy in an atomic inner shell induced by collisions with charged particles (see section 2.3.) or by photoionization [180], but also to the change of the nucleus charge in nuclear reactions such as nuclear beta decay, nuclear electron capture, internal conversion and alpha decay [181]. If the ionization probabilities can usually be calculated in the framework of the sudden approximation (SA), the accuracy of the calculation depends on how fast is the potential change compared with the relaxation time of the electrons in the new core potential. Nuclear beta decay offers ideal conditions since the potential change occurs in less than 10^{-18}s, which is the transit time of the emitted beta particles through the orbital electron cloud. Calculations of SO probabilities following beta decay were first performed using hydrogen like wave functions [182], and later, using more sophisticated numerical self-consistent wave functions for many electrons atoms [183]. However, the comparison between experiments and calculations is usually difficult because of secondary processes such as Auger electron emission that contribute to the final charge state of the daughter ions. Even for the simplest case investigated so far, the beta decay of ^{6}He atoms, the theoretical double ionization probability was overestimated by almost one order of magnitude [184].

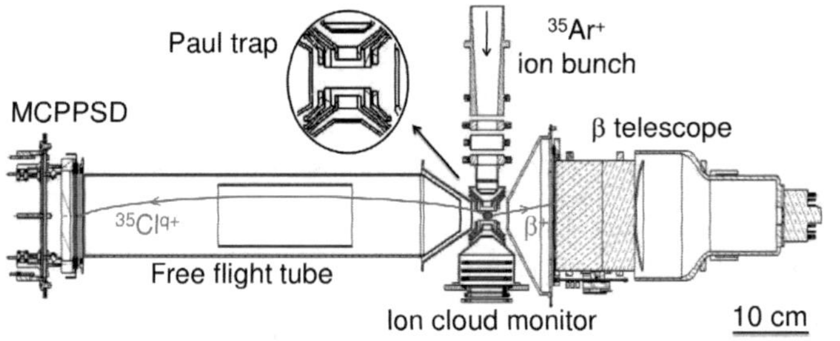

Figure 11. (Color online) Top view of the LPC trap setup and representation of a beta decay event from $^{35}Ar^+$ trapped ions. The recoiling daughter is accelerated between two collimators, enters a free flight region and is detected by a microchannel plate detector (MCPPSD). The detection of the beta particle provides a start for the recoil ion time of flight measurement, allowing charge state identification. The insert shows the structure of the Paul trap.

The LPC trap setup [185,186] has been initially developed to perform in-trap decay precision measurements [187] by confining radioactive ions from the SPIRAL target-ion source of GANIL in a transparent Paul trap. By implementing a new time-of-flight recoil ion spectrometer (figure 11) with the capability to separate the different charge states of the recoil ions resulting from beta decay, the setup has been recently used to investigate electron SO processes in the decay of singly charged radioactive ions. This new setup allowed, for the first time, to perform a SO measurement in the decay of $^6He^+$ ions. With only one single active electron, electron-electron correlations and secondary relaxation processes are absent, leaving only two possible mechanisms for the daughter ionization: direct collision with the beta particle, and the rapid change of both, the nuclear charge and the recoil velocity of the daughter nucleus. In this ideal textbook case, the experimental ionization probability was measured with a relative precision of the order of 10^{-4} and the result, P_{SO}=0.02339(36), was found in perfect agreement with simple mechanical calculations based on the SA [188].

Beyond the prototypical $^6He^+$case, heavier systems such as $^{35}Ar^+$ can reveal the role of more subtle shakeoff dynamics involving several electrons, and of subsequent relaxation processes such as the emission of Auger electrons. A recent joined experimental/theoretical study of the SO process in $^{35}Ar^+$ decay has provided a quite complete picture of ion formation [189] showing, in particular, the importance of Auger electrons. The same technique will be applied to $^{19}Ne^+$ ions in order to get a Z-dependent picture of the underlying ionization mechanisms.

5. Conclusions and perspectives

We have presented in the previous sections a large panel of collision experiments performed at GANIL during the three last decades, in the field of ion-atom and ion-molecule interactions. Covering a wide range of collision regime, these experiments led to a better understanding for several fundamental issues such as, for example, the role of electron-electron interaction in double capture, the importance of multiple-capture or-ionization processes depending on the collision regime or the limits of the two-step picture in ion-molecule collisions. New relaxation channels have also been evidenced, in particular for multi-ionized rare gas dimer targets with the identification of the ICD and RCT processes, and for OH^+ projectile ions leading unexpectedly to the formation of H^- ions. This experimental activity has given rise to several technological developments, such as the use of state-of-the-art X-ray spectrometers for the charge exchange process study, new spectrometers as well for the detection of molecular fragments, and the development of the recoil ion momentum spectroscopy technique. The quality and variety of the collected data have allowed testing and setting the limits of various theoretical models and calculations, such as the over-the-barrier model, the Landau-Zener model, CTMC calculations but also much more complex calculations such as close coupling based on atomic and molecular orbitals or two-center basis generator method. Beside the study of collisions, fundamental processes such as Young type interferences with a single electron or shakeoff ionization in nuclear beta decay are also part of the wide range of atomic physics topics that could be addressed at GANIL.

Most of these research topics are still very active, with a number of unanswered questions and a large place for new discoveries. We have seen that for medium and high energy collisions, the ICD process has a sizeable contribution in the emission of low energy electrons. This process can play an important role in radiation damage to biological tissues, and it is now necessary to extend its study to low energy collisions. Ion-molecule collisions have also been used as a tool to study Young type interferences with a single electron. Such collisions can be as well a powerful tool for molecular spectroscopy. As in the pioneering experiment performed with H_2^+ ions [190], collision-induced fragmentation of exotic molecules such as “trilobite” Rb_2 molecules [191] could yield capital information on the internuclear distance and on the molecule vibrational states. For the shakeoff ionization induced by nuclear reactions, few experimental data are presently available. Experiments involving radioactive ions with a wider range of atomic numbers would complete the present picture for this process. Finally, the future installation SPIRAL2 should open up an entirely new experimental field: the investigation of ion collisions with very dilute targets. In particular, the very high intensity of the high-energy ion beams that will be delivered to the S^3 spectrometer, coupled to a dedicated low energy platform equipped with an ECR source, will enable to study ion-ion collisions in the unknown intermediate velocity regime but where nevertheless the energy deposition is optimum. Besides the possibility to reach “pure” 3-body systems (bare ions on hydrogenic targets) as a benchmark to test theory in the simplest case, we will explore the role of additional electrons, bounded either to the target and/or to the projectile –one by one- to quantify the effects of the N-body quantum dynamics.

This project [192], apart from its fundamental interest, is also of prime importance for applications since it should provide a real breakthrough in the understanding of energy transfers in various plasmas (stellar /interstellar or inertial confusion plasmas) and also in the description, at the atomic level, of material modifications, including biological materials.

Acknowledgments
We would like to deeply thank the GANIL and CIMAP staff from the machine and the different experimental areas who allowed the realization of these experiments which often required ion beams of high optic quality. We are very grateful to all the students and postdoctoral fellows, coming from different countries, who were strongly involved in the whole of the work presented here. We are also deeply thankful to our numerous collaborators from France and to the international community who initiated and contributed to many experiments at GANIL in the field of atomic and molecular physics during the last three decades.

References
[1] Bransden B H and McDowell M R 1992 *Charge Exchange and the Theory of Ion-Atom collisions* (Oxford)
[2] Cravens T E 2002 *Science* **296** 1042
[3] Isler R C 1994 *Plasma Phys. Contr. F.* **36** 171
[4] Uehara S, Toburen L and Nikjoo H 2001 *Int. J. Radiat. Biol.***77** 139
[5] Uehara S and Nikjoo H 2002 *J. Phys. Chem.* B **106** 11051
[6] Ryufuku H, Sasaki K and Watanabe T 1980 *Phys. Rev.* A **21** 745
[7] Boduch P, Chantepie M, Hennecart D, Husson X, Kucal H, Lecler D, and Lesteven-Vaïsse I 1989 *J. Phys.* B **22** L377
[8] Jacquet E, Boduch P, Chantepie M, Druetta M, Hennecart D, Husson X, Lecler D, Olson R, Pascale J and Wilson M 1993 *Phys. Scripta* **47** 618
[9] Cowan R D 1981 *The theory of atomic structure and spectra* (Berkley : University of California Press)
[10] Boduch P, Chantepie M, Druetta M, Fawcett B, Hennecart D, Husson X, Kucal H, Stolterfoht N and Wilson M 1992 *Phys. Scripta* **45** 203
[11] Bazin V, Boduch P, Chantepie M, Cremer G, Jacquet E, Kucal H, Lecler D and Pascale P 2000 *Phys. Rev.* A **62** 052706
[12] Bazin V, Boduch P, Chantepie M, Jacquet E, Kucal H, Lecler D and Pascale P 2002 *Phys. Rev.* A **65** 032712
[13] Laulhé C, Jacquet E, Boduch P, Chantepie M, Ghérardi N, Husson X, Lecler D and Pascale J 1997 *J. Phys.* B **30** 2899
[14] Abrines R and Percival I C 1966 *Proc. Phys. Soc. Lond.* **88** 861
[15] Olson RE and Salop A 1977 *Phys. Rev.* A **16** 531
[16] Jacquet E, Pascale J, Boduch P, Chantepie M and Lecler D 1995 *J. Phys.* B **28** 2221
[17] Jacquet E, Chantepie M, Boduch P, Laulhé C, Lecler D and Pascale P 1999 *J. Phys.* B **32** 1151
[18] Fano U and Macek J H 1973 *Rev. Mod. Phys.* **45** 553
[19] Lin C D and Macek J H 1987 *Phys. Rev.* A **35** 5005
[20] Laulhé C, Jacquet E, Cremer G, Pascale J, Boduch P, Rieger G, Chantepie M and Lecler D 1997 *Phys. Rev.* A **55** 1088
[21] Jardin P *et al* 1993 Recoil Ion Kinetics *AIP Conf. Proc.* **274** (1993, New York : AIP) p 291
[22] Ullrich J *et al* 1994 *Comments At. Mol. Phys.* **30** 235
[23] Cocke C L *et al* 1991 *Phys. Rep.* **205** 153
[24] Cederquist H, Biedermann C, Selberg N and Hvelplund P 1995 *Phys. Rev.* A **51** 2169
[25] Cassimi A, Duponchel S, Fléchard X, Jardin P, Sortais P, Hennecart D and Olson R E 1996 *Phys. Rev. Lett.* **76** 3679

[26] Duponchel S, Cassimi A, Fléchard X, Hennecart D, Adoui L, Jardin P, Sortais P and Olson R 1997 *Phys. Scripta* **73** 199
[27] Zhang H *et al* 1999 *Phys. Rev.* A **60**, 3694
[28] Fléchard X, Harel C, Jouin H, Pons B, Adoui L, Frémont F, Cassimi A and Hennecart D 2001 *J. Phys.* B **34** 2759
[29] Roncin P, Barat M, Laurent H, Pommier J, Dousson S and Hitz D 1984 *J. Phys.* B **17** L521
[30] Stolterfoht N, Havener C C, Phaneuf R A, Swenson J K, Shafroth S M and Meyer F W 1986 *Phys. Rev. Lett.* **57** 74
[31] Barat M, Gaboriaud M N, Guillemot L, Roncin P, Laurent H and Andriamonje S 1987 *J. Phys.* B **20** 5771
[32] Harel C and Jouin H 1990 *Europhys. Lett.* **11** 121
[33] Harel C, Jouin H and Pons B 1991 *J. Phys.* B **24** L425
[34] Roncin P, Gaboriaud M N and Barat M 1991 *Europhys. Lett.* **16** 551
[35] Martin S, Denis A, Delon A, Désesquelles J and Ouerdane Y 1993 *Phys. Rev.* A **48** 1171
[36] Martin S, Bernard J, Denis A, Désesquelles J, Chen Li and Ouerdane Y 1994 *Phys. Rev.* A **50** 2322
[37] Martin S, Bernard J, Chen Li, Denis A and Désesquelles J 1995 *Phys. Rev.* A **52** 1218
[38] Frémont F, Sommer K, Hicham S, Boduch P, Lecler D and Stolterfoht N 1992 *Phys. Rev.* A **46** 222
[39] Frémont F, Sommer K, Hicham S, Boduch P, Lecler D and Stolterfoht N 1993 *Nucl. Instrum. Meth.* B **79** 3
[40] Frémont F, Merabet H, Chesnel J Y, Husson X, Lepoutre A, Lecler D, Rieger G and Stolterfoht N 1994 *Phys. Rev.* A **50** 3117
[41] Chesnel J Y, Merabet H, Husson X, Frémont F, Lecler D, Jouin H, Harel C and Stolterfoht N 1996 *Phys. Rev.* A **53** 2337
[42] Merabet H, Cremer G, Frémont F, Chesnel J Y and Stolterfoht N 1996 *Phys. Rev.* A **54** 372
[43] Chesnel J Y, Merabet H, Frémont F, Cremer G, Husson X, Lecler D, Rieger G, Spieler A, Grether M and Stolterfoht N 1996 *Phys. Rev.* A **53** 4198
[44] Chesnel J Y, Sulik B, Merabet H, Bedouet C, Frémont F, Husson X, Grether M, Spieler A and Stolterfoht N 1998 *Phys. Rev.* A **57** 3546
[45] Frémont F, Bedouet C, Husson X, and Chesnel J Y 1998 *Phys. Rev.* A **57** 4379
[46] Chesnel J Y, Merabet H, Sulik B, Frémont F, Bedouet C, Husson X, Grether M and Stolterfoht N 1998 *Phys. Rev.* A **58** 2935
[47] Chesnel J Y, Frémont F, Sulik B, Ruiz-Méndez C, Merabet H, Bedouet C, Husson X, Grether M and Stolterfoht N 1999 *Nucl. Instrum. Meth.* B **154** 142
[48] Frémont F *et al* 2002 *Int. J. Mol. Sci.* **3** 115
[49] Fléchard X, Duponchel S, Adoui L, Cassimi A, Roncin P and Hennecart D 1997 *J. Phys.* B **30** 3697
[50] Moretto-Capelle P, Bordenave-Montesquieu D and Bordenave-Montesquieu A 1999 *Phys. Scripta* **1999** 118
[51] Sanchez I and Bachau H 1995 *J. Phys.* B **28** 795
[52] Kazansky A K and Roncin P 1994 *J. Phys.* B **27** 5537
[53] Frémont F, Bedouet C, Chesnel J Y, Merabet H, Husson X, Grether M, Spieler A and Stolterfoht N 1996 *Phys. Rev.* A **54** R4609
[54] Bedouet C *et al* N 1999 *Phys. Rev.* A **59** 4399
[55] Barat M and Roncin P, 1992 *J. Phys.* B **25** 2205
[56] Mack M, Nijland J H, Straten P v d, Niehaus A and Morgenstern R 1989 *Phys. Rev.* A **39** 3846
[57] Vernhet D, Chetioui A, Rozet J P, Stephan C, Wohrer K, Touati A, Politis M F, Bouisset P, Hitz D and Dousson S 1988 *J. Phys.* B **21** 3949

[58] Martin S, Denis A, Ouerdane Y, Salmoun A, El Motassadeq A, Désesquelles J, Druetta M, Church D, and Lamy T 1990 *Phys. Rev. Lett.* **64** 2633
[59] Chetioui A *et al* 1990 *J. Phys.* B **23** 3659
[60] Stolterfoht N, DuBois R D and Rivarola R D 1997 *Springer Series in Atomic, Optical, and Plasma Physics* vol 20 (Berlin : Springer)
[61] Currel F J 2003 *The Physics of Multiply and Highly Charged Ions: Interactions with Matter* vol 2 (Dordrecht : Kluwer)
[62] Janev R K and Winter H 1985 *Phys. Rep.* **117** 265
[63] Ryufuku H, Sasaki K and Watanabe T 1980 *Phys. Rev.* A **21** 745
[64] Taulbjerg K 1986 *J. Phys.* B **19** L367
[65] Fritsch W and Lin C D 1984 *Phys. Rev.* A **29** 3039
[66] Winter T G and Lin C D 1984 *Phys. Rev.* A **29** 567
[67] Kimura M and Lin C D 1987 *Comments At. Mol. Phys.* **20** 35
[68] Errea L F, Gomez-Llorente J M, Mendez L and Riera A 1987 *J. Phys.* B **20** 6089
[69] Barany A, Astner G, Cederquist H, Danared H, Huldt S, Hvelplund P, Johnson A, Knudsen H, Liljeby L and Rensfelt K G 1985 *Nucl. Instrum. Meth.* B **9** 397
[70] Ali R, Cocke C L, Raphaelian M L A and Stockli M 1994 *Phys. Rev.* A **49** 3586
[71] Zhang H *et al* 2001 *Phys. Rev.* A **64** 012715
[72] Selberg N, Biederman C and Cederquist H 1996 *Phys. Rev.* A **54** 4127
[73] Beiersdorfer P, Olson R E, Brown G V, Chen H, Harris C L, Neill P A, Schweikhard L, Utter S B and Widmann K 2000 *Phys. Rev. Lett.* **85** 5090
[74] Tawara H, Richard P, Safronova U I and Stancil P C 2001 *Phys. Rev.* A **64** 042712
[75] Tawara H, Takacs E, Suta T, Makonyi K, Ratliff L P and Gillaspy J D 2006 *Phys. Rev.* A **73** 012704
[76] Allen F I, Biedermann C, Radtke R, Fussmann G and Fritzsche S 2008 *Phys. Rev.* A **78** 032705
[77] Cravens T E 2002 *Science* **296** 1042
[78] Krasnopolsky V A, Greenwood J B and Stancil P C 2004 *Space Sci. Rev.* **113** 271
[79] Otranto S and Olson R E 2011 *Phys. Rev.* A **83** 032710
[80] Trassinelli M, Prigent C, Lamour E, Mezdari F, Mérot J, Reuschl R, Rozet J-P, Steydli S and Vernhet D 2012 *J. Phys.* B **45** 085202
[81] Salehzadeh A and Kirchner T, 2013, *J. Phys.* B **49** 025201
[82] Salehzadeh A, Trassinelli M, Prigent C, Lamour E, Rozet J-P, Steydli S, Vernhet D, and Kirchner T 2014 *J. Phys. Conf. Ser.* **488** 082006
[83] Bethe H A 1930 *Ann. Physics* (*Leipzig*) **5** 325
[84] Stolterfoht N 1991 *Nucl. Instrum. Meth.* B **53** 477
[85] Crasemann B 1989 *Comments At. Mol. Phys.* **22** 163
[86] McGuire J H, Berrah N, Bartlett R J, Samson J A R, Tanis J A, Cocke C L and Schlachter A S 1995 *J. Phys.* B **28** 913
[87] Skogvall B *et al* 1995 *Phys. Rev.* A **51** R4321
[88] Stolterfoht N *et al* 1998 *Phys. Rev. Lett.* **80** 4649
[89] Stolterfoht N *et al* 1999 *Phys. Rev.* A **59** 1262
[90] Stolterfoht N *et al* 2000 *AIP Conf. Proc.* **506** 427
[91] Tanis J A *et al* 1999 *Phys. Scripta* **T80** 381
[92] Tanis J A *et al* 1999 *Phys. Rev. Lett.* **83** 1131
[93] Tanis J A *et al* 2000 *Phys. Rev.* A **62** 032715
[94] Chesnel J Y *et al* 2002 *Photonic, Electronic and Atomic collisions* ed Burgdörfer J et al (*Rinton Press Princeton*) p 699
[95] Tanis J A *et al* 1998 *Phys. Rev.* A **57** R3154
[96] Rangama J, Hennecart D, Stolterfoht N, Tanis J A, Sulik B, Frémont F, Husson X and Chesnel J Y 2003 *Phys. Rev.* A **68** 040701(R)

[97] Bapat B *et al* 1999 *J. Phys.* B **32** 1859
[98] Schulz M *et al* 2001 *J. Phys.* B **34** L305
[99] Schulz M *et al* 2003 *Nature* **422** 48
[100] Lüdde H J *et al* 1995 *J. Phys.* B **28** 4101
[101] Kirchner T, Lüdde H J, Hortabsch M 2004 *Phys. Scripta* **T110** 364
[102] Montenegro E C, Meyerhoff W E, McGuire J H 1994 *Adv. Atom. Mol. Opt. Phys.* **34** 249
[103] Montenegro E C, Santos A C F, Melo W S, Sant'Anna M M and Sigaud G M 2001 *Phys. Rev. Lett.* **88** 013201
[104] Montanari C, Miraglia J E, Arista N R 2003 *Phys. Rev.* A **67** 062702
[105] Kirchner T et al 2005 *Phys. Rev.* A **72** 012707
[106] Monti J M *et al* 2011 *J. Phys.* B **44** 195206
[107] Fregenal D *et al* 2014 *J. Phys. Conf. Ser.* **488** 082014
[108] Chabot M *et al* 1994 *J. Phys.* B **27**, 111
[109] Adoui L *et al* 1994 *Nucl. Instrum. Meth.* B **87**, 45
[110] Adoui L *et al* 1994 *Nucl. Instrum. Meth.* B **98**, 312
[111] Vernhet D *et al* 1996 *Nucl.Instrum. Meth.* B **107**, 71
[112] Vernhet D *et al* 1997 *Phys. Rev. Lett.* **83** 1131
[113] Ullrich J *et al* 2000 *J. Phys.* B **30** 2917
[114] Dörner R *et al* 2000 *Phys. Rep.* **330** 95
[115] Cassimi A 1999 *Phys. Scripta* **T80** 98
[116] Jardin P *et al* 1996 *Nucl.Instrum. Meth.* B **107** 41
[117] Shah M B and Gilbody H B 1990 *J. Phys.* B **23** 1491
[118] McLaughlin T K, McCullough R W and Gilbody H B 1992 *J. Phys.* B **25** 1257
[119] Meng L, Olson R E, Folkerts H O and Hoekstra R 1994 *J. Phys.* B . **27** 2269
[120] Folkerts H O, Hoekstra R and Morgenstern R 1996 *Phys. Rev. Lett.*77 3339
[121] Ben-Itzhak I, Wells E, Stöckli M P, Tawara H and Carnes K D 1997 *Phys. Scripta* **T73** 270
[122] Tarisien M, Adoui L, Frémont F and Cassimi A 1999 *Phys. Scripta* **T80** 182
[123] Wood C J and Olson R E 1999 *Phys. Rev.* A **59** 1317
[124] DuBois R D, Ali I, Cocke C L, Feeler C R and Olson R E 2000 *Phys. Rev.* A **62** 060701R
[125] Tarisien M *et al* 2000 *J. Phys.* B **33** L11
[126] Frémont F, Bedouet C, Tarisien M, Adoui L, Cassimi A, Dubois A, Chesnel J Y and Husson X 2000 *J. Phys.* B **33**, L249
[127] Adoui L *et al* 2001 *Phys. Scripta* **T92** 89
[128] Frémont F, Bedouet C, Husson X, Chesnel J Y and Stolterfoht N 1999 *Phys. Rev.* A **60** 3727
[129] Sobocinski P, Rangama J, Chesnel J Y, Tarisien M, Adoui L, Cassimi A, Husson X and Frémont F 2001 *J. Phys.* B **34** L367
[130] Sobocinski P, Rangama J, Chesnel J Y, Tarisien M, Adoui L, Cassimi A, Husson X and Frémont F 2001 *AIP Conf. Proc.* **576** 114
[131] Sobocinski P, Rangama J, Laurent G, Adoui L, Cassimi A, Chesnel J Y, Dubois A, Hennecart D, Husson X and Frémont F 2002 *J. Phys.* B **35** 1353
[132] Sobocinski P *et al* 2003 *J. Phys.* B **36** 1283
[133] Frémont F, Martina D, Kamalou O, Sobocinski P, Chesnel J Y, McNab I R and Bennett F R 2005 *Phys. Rev.* A **71** 042706
[134] Juhász Z, Chesnel J Y, Frémont F, Hajaji A and Sulik B 2008 *AIP Conf. Proc.* **1080** 118
[135] Juhász Z, Sulik B, Frémont F, Hajaji A and Chesnel J Y 2009 *Nucl. Instrum. Meth.* B **267** 326
[136] Sobocinski P, Pešić Z D, Stolterfoht N,Sulik B, Legendre S and Chesnel J Y 2005 *J. Phys.* B **38** 2495
[137] Sobocinski P, Pešić Z D, Hellhammer R, Klein D,Sulik B, Chesnel J Y and Stolterfoht N 2006 *J. Phys.* B **39** 927

[138] Juhász Z, Frankland B S, Frémont F, Rangama J, Chesnel J Y and Sulik B 2012 *J. Phys. Conf. Ser.* **388** 102051
[139] Juhász Z, Sulik B, Rangama J, Bene E, Sorgunlu-Frankland B, Frémont F and Chesnel J Y 2013 *Phys. Rev.* A **87** 032718
[140] Lattouf E *et al* 2014 *Phys. Rev.* A **89** 062721
[141] Cassimi A *et al* 2003 Many-particle quantum dynamics in atomic and molecular fragmentation *Atomic, Optical and Plasma Physics* (Springer-Verlag Berlin Heidelberg New-York) ed J Ullrich and V P Shevelko pp 1615-565
[142] Tarisien M *et al* 2000 *J. Phys.* B **33** L11
[143] Wohrer K and Watson R L 1993 *Phys. Rev.* A **48** 4784
[144] Caraby C, Cassimi A, Adoui L and Grandin J P 1997 *Phys. Rev.* A **55** 2450
[145] Kabachnik N M, Kondratyev V N, Roller-Lutz Z and Lutz H O 1998 *Phys. Rev.* A **57** 990
[146] Kabachnik N M, Kondratyev V N, Roller-Lutz Z and Lutz H O 1997 *Phys. Rev.* A **56** 2848
[147] Legendre S *et al 2005 J. Phys.* B **38** L233
[148] Muranaka T 2007 PhD thesis Univeisté de Caen Basse-Normandie
[149] Morishita Y *et al* 2006 *Phys. Rev. Lett.* **96** 243402
[150] Havermeier *T et al* 2010 *Phys. Rev. Lett.* **104** 133401
[151] Cederbaum L S, Zobeley J and Tarantelli F 1997 *Phys. Rev. Lett.* **79** 4778
[152] Jahnke T *et al* 2010 *Nature Phys.* **6** 139
[153] Averbukh V *et al* 2011 *J. Electron Spectrosc.* **183** 36
[154] Kim H K *et al* 2013 *Phys. Rev.* A **88** 042707
[155] Ben-Itzhak I, Ginther S G and Carnes K D 1993 *Phys. Rev.* A **47** 2827
[156] Wohrer K, Sampoll G, Watson R L, Chabot M, Heber O and Horvat V 1992 *Phys. Rev.* A **46** 3929
[157] Remscheid A, Huber B A, Pykavyj M, Staemmler V and Wiesemann K 1996 *J. Phys.* B **29** 515
[158] Matsumoto J *et al* 2010 *Phys. Rev. Lett.* **105** 263202
[159] Matsumoto J *et a*l 2011 *Phys. Scripta* **T144** 014016
[160] Iskandar W *et al* 2014 *Phys. Rev. Lett.* **113** 143201
[161] Stolterfoht N *et al* 2001 *Phys. Rev. Lett.* **87** 023201
[162] Stolterfoht N *et al* 2003 *Phys. Rev.* A **67** 030702R
[163] Stolterfoht N, Sulik B, Skogvall B, Chesnel J Y, Frémont F, Hennecart D, Cassimi A, Adoui L, Hossain S, and Tanis J A 2004 *Phys. Rev.* A **69** 012701
[164] Sulik B, Tanis J A, Chesnel J Y and Stolterfoht N 2004 *Phys. Scripta* **T110** 345
[165] Frémont F Hajaji A, Naja A, Leclercq C, Soret J, Tanis J A, Sulik B, Chesnel J Y 2005 *Phys. Rev.* A **72** 050704
[166] Tanis J A, Chesnel J Y, Sulik B, Skogvall B, Sobocinski P, Cassimi A, Grandin J P, Adoui L, Hennecart D, and Stolterfoht N 2006 *Phys. Rev.* A **74** 022707
[167] Chesnel J Y,Hajaji A, Barrachina R O and Frémont F 2007 *Phys. Rev. Lett.* **98** 100403
[168] Chesnel J Y,Hajaji A, Barrachina R O and Frémont F 2007 *J. Phys. Conf. Ser.* **58** 185
[169] Frémont F, Hajaji A, Barrachina R O and Chesnel J Y 2007 *J. Phys. Conf. Ser.* **88** 012020
[170] Frémont F, Hajaji A, Barrachina R O and Chesnel J Y 2007 *Reflets de la Physique, Journal SFP* **10** 10
[171] Frémont F, Hajaji A, Barrachina R O and Chesnel J Y 2008 *CR Phys.* **9** 469
[172] Frémont F, Suarez S, Barrachina R O, Hajaji A, Sisourat N, Dubois A and Chesnel J Y 2009 *Nucl. Instrum. Meth.* B **267** 206
[173] Frémont F, Gruyer D, Helaine V, Fossez K, Leredde A, Maclot S, Lepailleur A, Hajaji A, Chesnel J Y, Barrachina R O and S. Suárez 2009 *Handbook of Interferometers; Research, Technology and Applications* (Nova Science Publishers, Hauppauge, NY) ed D Halsey and W Raynor

[174] Barrachina R O, Frémont F, Fossez K, Gruyer D, Helaine V, Lepailleur A, Leredde A, Maclot S, Scamps G, and Chesnel J Y 2010 *Phys. Rev.* A **81** 060702R
[175] Frankland B, Barrachina R O, Chesnel J Y, Frémont F 2012 *J. Phys. Conf. Ser.* **388** 102009
[176] Vabre M, Girard S, Gilles H, Frankland B S, Porée F, Leprince P, Chesnel J Y, Barrachina R O, and Frémont 2012 *ISRN Spectroscopy* **2012** 174952
[177] Frémont F 2012 *Interférences électroniques Physique atomique - Interférences macroscopiques et nanoscopiques- Analogies avec les interférences photoniques* (Ellipses, Technosup, 2012)
[178] Frémont F 2013 *Springer Series on Atomic, Optical, and Plasma Physics* vol 77 (Springer, 2013)
[179] Kamalou O, Chesnel J Y, Martina D, J. Hanssen, Stia CR, Fojón O A, Rivarola R D, and Frémont F 2005 *Phys. Rev.* A **71** 010702R
[180] Wehlitz R, Heiser F, Hemmers O, Langer B, Menzel A and Becker U 1991 *Phys. Rev. Lett.* **67** 3764
[181] Freedman M 1974 *Annu. Rev. Nucl. Sci.* **24** 209
[182] Levinger JS 1953 *Phys. Rev.* **90** 11
[183] Carlson T A, Nestor C W, and Tucker J C 1968 *Phys. Rev.* **69** 27
[184] Wauters L and Vaeck N 1996 *Phys. Rev. C* **53** 497
[185] Fléchard X *et al* 2008 *Phys. Rev. Lett.* **101** 212504
[186] Rodriguez D *et al* 2006 *Nucl. Instrum. Meth.* A **565** 876
[187] Fléchard X *et al* 2011 *J. Phys.* G **38** 055101
[188] Couratin *C et al* 2012 *Phys. Rev. Lett.* **108** 243201
[189] Couratin *C et al* 2013 *Phys. Rev.* A **88** 041403 R
[190] Shmidt L Ph H, Jahnke T, Czasch A, Schöffler M, Schmidt-Böcking H and Dörner R 2012 *Phys. Rev. Lett.* **108** 073202
[191] Greene C H, Dickinson A S and Sadeghpour H R 2000 *Phys. Rev. Lett.* **85** 2458
[192] *http://www.insp.jussieu.fr/Collisions-Ion-Lent-Ion-Rapide.html*

Journal of Physics: Conference Series **629** (2015) 012002 doi:10.1088/1742-6596/629/1/012002

Transport of fast excited ions through thin solids probed by X-ray spectroscopy

E Lamour[1,2], J-P Rozet[1,2], D Vernhet[1,2], B Gervais[1], J-P Grandin[1], D Lelievre[1], J-M Ramillon[1] and H Rothard[1]

[1] CNRS, UMR 7588, Institut des NanoSciences de Paris (INSP), F-75005 Paris, France

[2] Sorbonne Universités, UPMC Univ. Paris 06, UMR 7588, INSP, F-75005 Paris, France

[3] Centre de Recherche sur les Ions, les Matériaux et la Photonique (CIMAP), UMR 6252 (CEA/CNRS/ENSICAEN/UCBN), F-14070 Caen Cedex 05

E-mail: emily.lamour@upmc.fr

Abstract. Over more than a decade, experimental studies of the production and transport of projectile excited states in thin solid targets have been performed at GANIL for Ar^{17+} and Kr^{35+} in the so-called high velocity domain. A range of target thicknesses from single collision condition to equilibrium has been investigated. X-ray spectroscopy techniques have allowed us to determine *absolute* $n\ell$ populations of core and Rydberg projectile states, as well as the relative population of fine structure substates ($n\ell_j$). In parallel, theoretical simulations to treat on the same footing all the competing processes, i.e., collisions, radiative decay and dynamical screening due to the wake field, have been developed. Methods based on either master equations or Monte Carlo approaches have allowed us to reach an unprecedented precision in the description of the ion transport in matter in the perturbative regime. In particular, a first direct measurement of the wake field usually extracted from ion stopping power has been performed.

1. Introduction

Fast Highly Charged Ions (HCIs) going through matter induce material modifications like, for instance, the formation of tracks in insulating materials, and in turn, their stopping power, charge state and excited states depend on the material encountered. Two different representations have been used to describe the interaction between fast highly charged ions and solid targets. They correspond to different approaches for calculating, for instance, the slowing down of a projectile through materials. In the first picture, one assumes that the ion-solid interaction is the result of a series of binary collisions with the target electrons, and *ion-atom cross sections* correspond to the data base of the *collisional* calculations. For the ion stopping power, this approach is directly related to the Bethe theory [1] taking into account explicitly the distortion of target wave-functions (more details can be found in [2] and [3]). In the other picture, closer in spirit to the *dielectric* theory first proposed by Bohr [4], the target electrons are considered to *respond collectively* to the passage of the projectile. The HCIs induce a polarization of the medium described as a wake of electronic density fluctuation trailing the ion, the so-called *wake field.* The gradient of the wake potential leads to the establishment of an electric field, and its local value at the projectile can be used to calculate the stopping power [5-7].

Journal of Physics: Conference Series **629** (2015) 012002 doi:10.1088/1742-6596/629/1/012002

Both types of calculations appear to achieve good agreement with stopping power measurements, despite a rather different physical picture for the behavior of the target electrons.
The purpose of our work shortly presented here, was to evaluate the role played by target electrons in these two pictures and to define eventually the validity limits and the applicability of corresponding models in terms of different experimental observables. It is clear that the response of the target electrons to the passage of the ion has direct consequences on the projectile ion: it is not only slowed down, but, the population of its excited states is also altered by the surrounding medium. Notably, the presence of the wake field may mix the ion excited states by Stark effect in a coherent manner while a pure collisional response of target electrons destroys any coherence. Hence, we tackled studies on the production and transport of projectile excited states by analyzing the de-excitation of these excited states to probe the solid response. The theoretical description of the ion-solid interaction has been a real challenge. Indeed, the evolution of atomic projectile states propagating through a solid originates from both dissipation and dephasing. It turned out that to address such a problem the best is to consider the ion projectile as an open quantum system (the excited ion) coupled to reservoirs representing the solid and the radiation field responsible for photon emission. We developed several complementary and overlapping types of theoretical methods: a classical simulation based on a stochastic approach of the ion transport [5] and two quantum simulations implying either the resolution of the stochastic time-dependent Schrödinger equation [6, 7] or of master equations [8, 9] as "rate"-like equations.
Several experimental observables were determined for different collision systems in the perturbative regime. We remind that the perturbative regime is reached for high projectile velocities (projectile velocity much higher than the velocity of the active electron) and/or for $Z_p \gg Z_t$ (where Z_p and Z_t stand for atomic numbers for the projectile and the target, respectively). In this regime, ionization and excitation have cross sections much larger than capture. In this publication, we restrict the presentation to a selection of our results classified in order of increasing sensitivity to either the individual processes involved and/or to the dynamical screening due to the wake field. Accordingly, we first present the populations of Rydberg states, we then discuss the core $n\ell$ (with $n < 5$) states, and finally we examine the evolution of fine structure sub-state populations ($n\ell_j$ with $n = 2$ or 3 depending on the ion species). In the next section, we give the main characteristics of the experimental methods and describe the techniques applied to determine each observable. Section 3 briefly presents the theoretical treatments. Section 4 is dedicated to an overview of the results and a comparison with the simulations. Our conclusions are summarized in the last section.

2. Experiments

All the experiments have been performed with the GANIL facilities in Caen (France) using either SME (Sortie Moyenne Energie) or LISE (Ligne d'Ions Super Epluchés) experimental areas. These beam lines allow us to work with high optical quality beams from the intermediate to the high energy regime.

2.1. Collision systems

The choice of collision systems, i.e., the ion projectile, the collision velocity, the nature of the solid and its thickness was guided to highlight the various transport effects. For the $n\ell$ populations of projectile excited states, the significance of the results arises from a simplification of the problem. To provide stringent benchmarks for theory, experiments were performed with:

- the "primary" production process of the excited states well identified and the single collision condition fulfilled even for the thickest target thickness (meaning that the multiple process probability is at most a few percent); the results, selected here, concern either capture process or single excitation processes.
- the transport of the excited states –mainly due to excitation and ionization processes- investigated over a wide range of collision numbers, from single collision conditions to the equilibrium of populations.

These conditions imply to be in the perturbative regime and to study hydrogen-like excited states for which the electronic structure and the coupling between the excited states are well known. Here, we report on major results obtained for excited Kr^{35+} (33.2 or 60 MeV/u) and Ar^{17+} (13.6 MeV/u) ions with carbon targets. Other targets have been used, like copper or aluminum [12, 10]. The target thickness ranges from 100 Å to 1 μm. As a consequence, the ion transit time inside the target is at maximum a few tens of femtoseconds. This time is long enough for the excited states evolution to be sensitive to transport effects. Indeed, the characteristic time for the establishment of the wake field (i.e., the wake period related to the plasma frequency) is less than 0.2 fs and the Stark mixing time, for the excited configurations under investigations, remains smaller than 9 fs and than 3 fs in the case of krypton and argon, respectively. Moreover, a significant Stark effect may be observed since the radiative lifetime of those excited states is larger than the Stark mixing time for $n \geq 3$ for krypton and $n \geq 2$ for argon. Finally, the mean free times of collisional processes depend on the nature of the process and decrease rapidly with the value of the principal quantum number but its contribution can be explored also with the target thickness.

2.2. Techniques

We performed X-ray spectroscopy measurements that allow determining the population of projectile excited states. The typical experimental set-up is schematically represented Figure 1. HCIs are directed onto self-supported carbon foils with measured thicknesses and purity. In some cases, gaseous targets have also been used to control the initial population produced by the primary process. Projectile X-ray emission is recorded by different detection systems depending on the observable we want to determine:

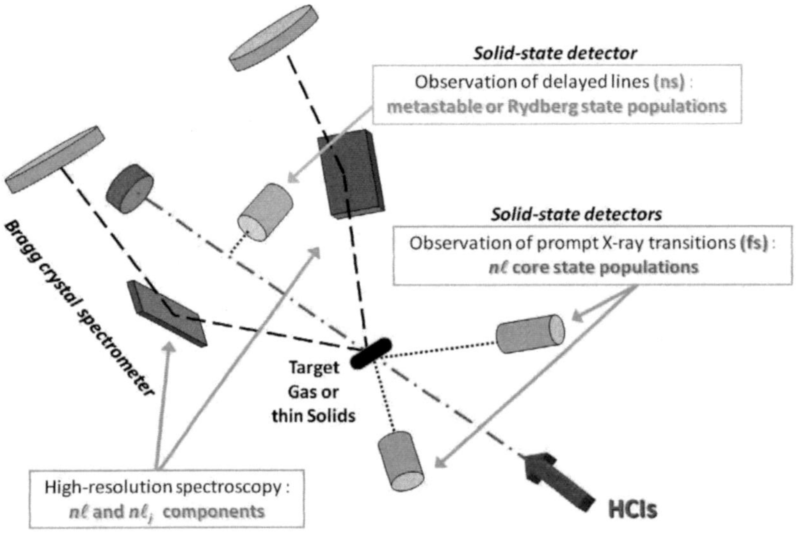

Figure 1. Schematic representation of the complete experimental set-up.

- The study of *the delayed photon emission* of short lifetime excited states (as 2p or 3p for hydrogen-like ions) is an important tool for investigating Rydberg states emerging from the solid. We have examined the delayed emission of exited Ar^{17+} ions, for each target thickness, by means of a solid-state detector placed at 90° with respect to the beam direction. Delay times, behind the target, up to 1 ns are explored with a resolution of 6×10^{-12} s. The $2s_{1/2}$ metastable state population is also determined using this system by the observation of the two-photon (2E1) transition (see figure 2).

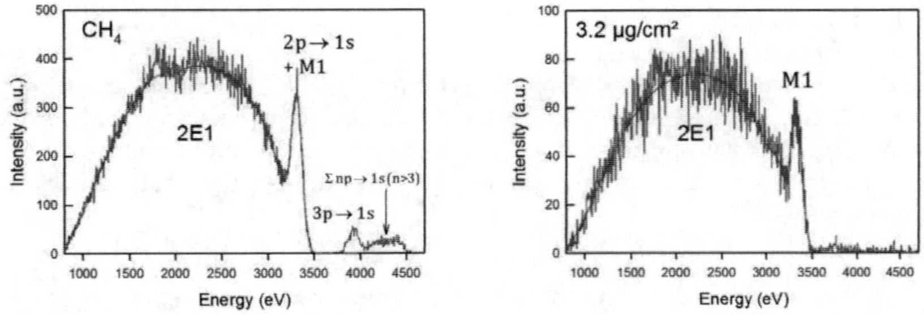

Figure 2. Spectra of the delayed X-ray transitions emitted by Ar^{17+} ions at 13.6 MeV and recorded by a solid-state detector placed at D = 127 mm behind the target for a CH_4 gaseous target and a 3.2 μg/cm² carbon foil. The 2E1 decay mode as well as the M1 one from the de-excitation of the 2s state are clearly distinguishable. Note: the photon energy is given in the projectile frame.

- *Prompt Lyman transition* intensities are measured also with solid-state detectors recording the emitted (3 to 5 keV) X-rays with the target placed at the center of the collision chamber. This allows to obtain the evolution of *np* core state populations as a function of target thickness since a very small fraction (≤ 3%) of the projectile excited states lies above $n = 10$.

For each type of emission lines – *delayed or prompt-* the energy resolution of the solid-state detectors allows us to identify Lyman lines with n ≤ 4 (as well as the 2E1 decay mode of the 2s metastable state) in the case of Ar ions and for n ≤ 5 in the case of Kr. The detection efficiency has been carefully determined through dedicated experiments [11].

- The effect of transport processes on *fine structure components* of core excited states is determined using high-resolution high-transmission Bragg-crystal spectrometers equipped with mosaic graphite crystals (HOPG) and position sensitive detectors. Our crystal spectrometers are placed in a vertical geometry, allowing first order correction of the broadening of the emitted lines due to Doppler effect. A detailed description is presented in [12]. Additionally to the fine structure components, in the case of argon ions, we were able to record, with an extremely high resolution, the complete prompt Lyman series as shown in Figure 3.

Each detection system is placed at a specific angle with respect to the beam direction to be polarization-insensitive. Spectra recorded in the case of krypton ions are shown in [12] and [13].

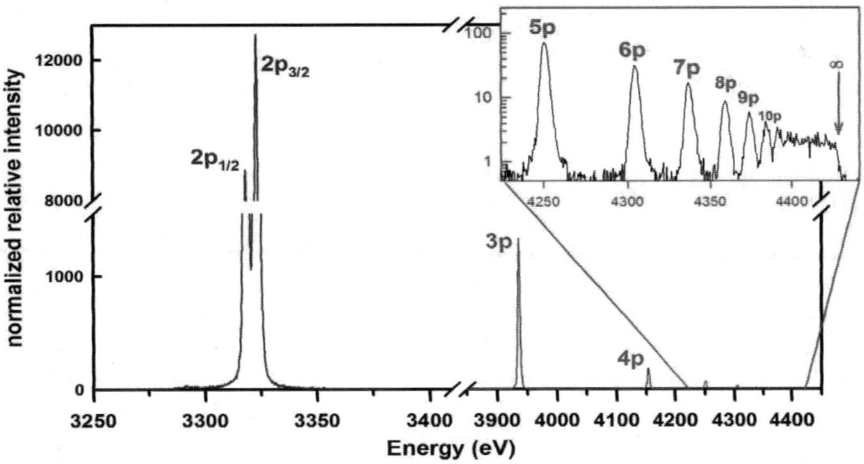

Figure 3. High resolution spectra of H-like X-ray transitions produced during the collision of Ar^{18+} ions at 13.6 MeV with a 47.2 μg/cm² (i.e. 220 nm) carbon target. All the np → 1s lines with n ∈ [2, ∞] are recorded by accumulating data over different Bragg angle settings of the spectrometer. For the 2p level, the two components $2p_j$ with j = 1/2 and 3/2 are resolved. Note: the photon energy is given in the projectile frame.

Journal of Physics: Conference Series **629** (2015) 012002 doi:10.1088/1742-6596/629/1/012002

3. Theoretical approaches

The ion-solid interaction represents a good example of a small open system in contact with a large reservoir. The open system is the internal state of a fast hydrogen-like ion while the reservoir consists of the radiation field and of the particles of the solid that induce energy exchange and decoherence in the small system. Hence, the starting point of any description of the ion-solid interaction is the reduced density operator $\sigma(t)$ whose time evolution follows the Liouville equation that includes dissipative and non-dissipative terms:

$$i\frac{\partial}{\partial t}\sigma(t) = [H_S, \sigma(t)] + R\sigma(t) \qquad (1)$$

where H_S is the Hamiltonian of the small system which contains the Coulomb interaction potential possibly modified by the wake potential and R the relaxation operator describing its interaction with the reservoir.

We have suggested several theoretical simulations to solve equation (1) using different methods and various approximations. The coupling between the small open system and the reservoir is perturbative. Consequently, we show that the coupling element between two states due to the wake field is much smaller than the inverse of correlation times of vacuum fluctuations on one hand and of collisional processes on the other hand. It is then possible to use an extension of Bloch equations (see the work of C. Cohen-Tanoudji, J. Dupont-Roc and G. Grynberg [13]) or, in other words, a *master equations approach* (MEA) ([8, 11, 12]). This first approach consists of solving a set of differential coupled equations that allows following the evolution of populations of each $n\ell j m_j$ state and of coherences between states. In our approach, we neglect the damping of coherences through collisions. All the transition rates are calculated explicitly in the hydrogenic basis set used to describe the projectile electronic states. We account for all the cross sections of processes involved during transport (in fact, for collision, atomic cross sections of capture, ionization, intra-shell excitation with $\Delta\ell = \pm 1$ and ± 2 and inter-shell excitation with $\Delta n = \pm 1$ and ± 2 are included). In this treatment, the basis set consists of all the sub-states of the n = 1–6 shells. Actually, we found that convergence of calculations is reached around n = 6 (changes when going from n = 5 to 6 do not exceeds a few %). When neglecting the couplings between the off-diagonal matrix elements (the coherence effects) and the diagonal matrix elements of the reduced density matrix, the problem reduces simply to the resolution of differential equations giving the time evolution of only the excited $n\ell$ state populations referred to as *rate equations*. The *quantum transport theory* (QTT) initiated by the groups of J. Burgdörfer and C.O. Reinhold [9, 10, 14, 15] has been developed to overcome the problem of coherence transport and the limit in the number of states to be considered. The starting point of such a description is the possibility to write the relaxation operator R in equation (1) using the Lindblad approach [10, 16]. This insures the possibility to solve the quantum Liouville equation by means of a quantum trajectory Monte Carlo technique first used in quantum optics for the description of few-state atomic systems interacting with the radiation field (see for example the work of J. Dalibard *et al* [17]). In this approach, the reduced density operator $\sigma(t)$ of the system is constructed from the independent evolution of an ensemble of N_{traj} pure states. Each pure state evolution is called a quantum trajectory and describes a different random sequence of interactions with the environment through the non-linear stochastic Schrödinger equation. The QTT method can be considered as a quantized version of the corresponding *classical transport theory* (CTT) for which test particles (electron of the projectile ion) obey then the Langevin equation. This equation describes the motion of the electron on a classical orbit disturbed by a stochastic force [21-23]. This classical method does not take into account for the effects related to the spin of the electron and the radiative decay. As a result, the classical approximation will be only valid for Z_p ions not too heavy for which the spin-orbit coupling is rather weak and the radiative decay times long enough compared to the ion transit time inside the solid. These either deterministic or stochastic approaches have been compared between each other [9, 18] showing that the quantum MEA and QTT methods give similar results. What is interesting is the possibility to gauge the sensitivity of the experimental data to the transport effects by, for instance, switching "on/off" the influence of the wake field. Additionally, we may test the initial density matrix representing the primary process of population and notably the initial phase and extract an effective wake field value.

4. Results and discussions

One of our major contributions to the topics treated here is to have been able to confront theoretical approaches with experimental data on an absolute scale. Great cares have been taken to extract absolute experimental cross sections and, in parallel, efforts have been made to include in the theoretical developments all the ingredients so as to treat the complete dynamics of the interaction and therefore to reach a real understanding in the relevant parameters.

4.1. Rydberg state populations

An exhaustive experimental study of the production and transport of Rydberg excited states has been done for Ar^{18+} ions on solid carbon targets, at a velocity of 23 a.u., and for a range of thicknesses allowing us to span the transport conditions from single collision to equilibrium (3.5 to 201 μg/cm²). In this case, the number of collisions the projectile electron experiences in the solid is extensively varied from ~ 0.6 to 60 collisions in average. The evolution of the Lyα(*2p→1s*) and Lyβ(*3p→1s*) intensities as a function of the ion time of flight behind the thinnest and the thickest carbon foils used in the experiment is shown in Figure 4 (a series of results for the other target thicknesses are given in [8]). Keeping in mind that lifetimes of hydrogen-like $n\ell$ states are proportional to $n^3\ell^2$ for a given projectile, the delayed photon emission observed in the decay of *2p* and *3p* states comes from the Rydberg state population through cascade contributions. The observed *2p* decay time is much longer than the *3p* one for all target thicknesses, which is indeed a direct signature of high angular momentum Rydberg states. With the CTT model, we demonstrate that the *2p* state population, for this system, is sensitive to excited states up to n = 30 even though only 3% of the excited states have a principal quantum number $n > 10$. The comparison with the experimental data shows that the evolution of the population of highly excited states during the transport is mainly governed by multiple scattering (intra and inter-shell excitation processes). Indeed, the classical simulation with or without the wake field does not exhibit significant differences.

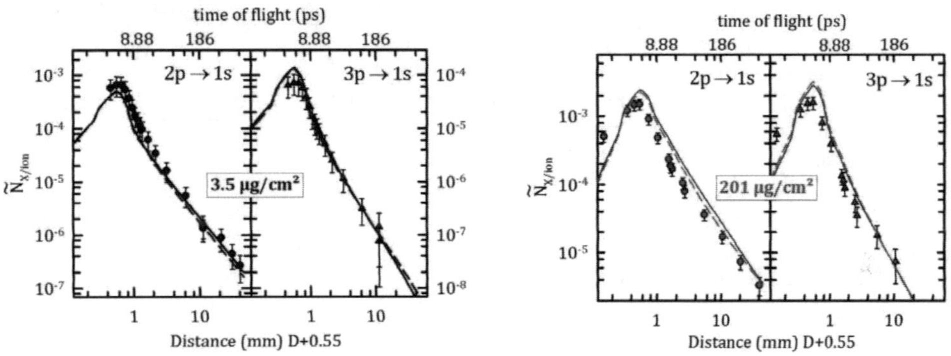

Figure 4. Evolution with the ion time of flight behind the target of the Ar^{17+} Lyman line intensities, i.e., number of emitted photons per ion, for the 3.5 and 201 μg/cm² carbon target [8]. Experimental results: symbols, CTT simulations with wake "off" (dashed lines) and wake "on" (solid lines). In the simulations, Continuum Distorted Wave (CDW) calculations have been used to account for the primary capture processes. The distance behind the target has been arbitrarily shifted by 0.55 mm for sake of clarity.

4.2. Core state populations

We illustrate in Figure 5 the evolution of absolute core *np* state populations as a function of target thickness, or alternatively as a function of ion transit time through solid foils for Kr^{35+} at 33.2 MeV/u and Ar^{17+} at 13.6 MeV/u. The dotted lines reported in Figure 5 that account only for binary ion-atom collisions i.e., without any transport effects, highlight clearly that the core *np* populations are sensitive to transport effects, especially, when increasing the principal quantum number. When comparing with the dashed lines, we observe that the major contribution in the evolution of the experimental data is due to collisional processes and radiative decay (note that in the case of argon only 3% of states deexcite inside the solid). The sensitivity to the wake field is weak and appears more clearly in the case of argon mainly because the wake field scales roughly as $1/v_p^2$ with v_p the projectile velocity. Similar behaviors are observed for the evolution the *np* Lyman lines of Kr^{35+} at 60 MeV/u populated by the single 1s→2p excitation process as primary process [24]. In the case of argon, the 2s state population has also been examined and the great sensitivity to transport effects for the core state are shown in figure 6.

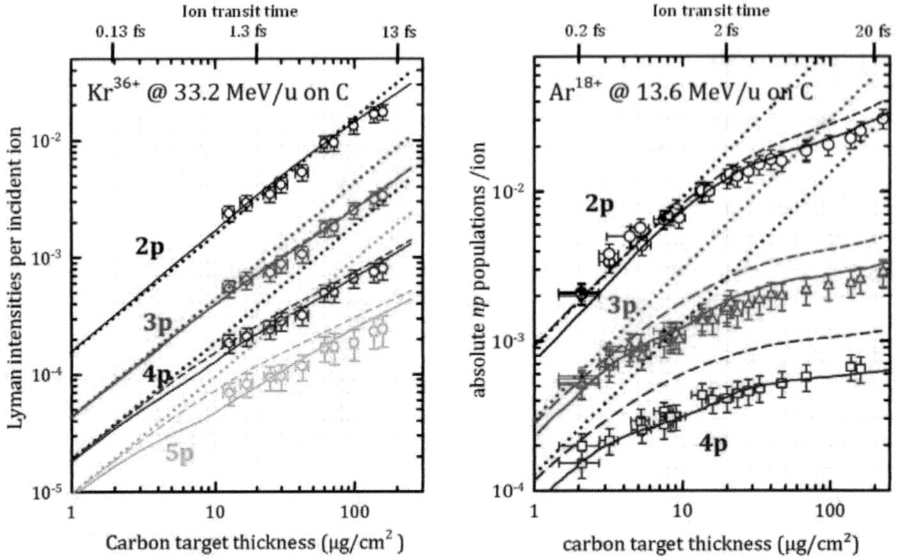

Figure 5. Absolute Lyman intensities (or populations) as a function of carbon target thickness (i.e., ion transit time) for two different collision systems. The dotted lines account only for binary ion-atom collisions (i.e., without any transport effects). The full and dashed lines correspond to the predictions of quantum approach (MEA for krypton and QTT for argon) with "wake on" and "wake off" respectively. In both cases, CDW calculations are used for the initial conditions of populations, i.e., the capture process.

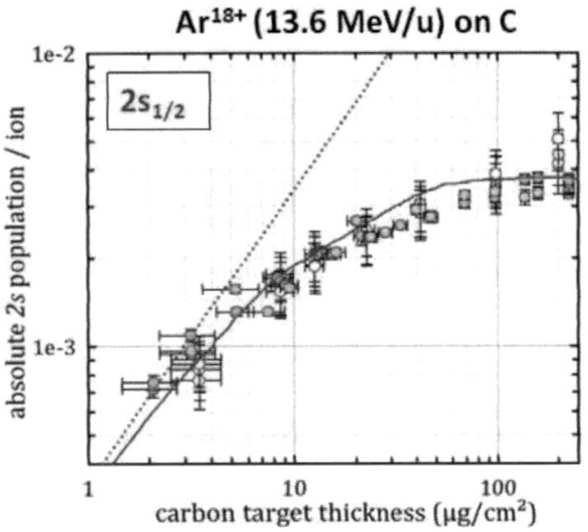

Figure 6. Evolution the 2s population as a function of the foil thickness. Symbols: experiment data. Dotted green line: results when no transport effects are included. Full green line: results predicted by QTT taking into account all the transport effects including the wake field [9].

4.3. Fine structure sub-state populations of core states

The observable exhibiting a strong sensitivity to the wake field is the evolution of the fine structure $n\ell j$ components, which is related to the Stark mixing and, consequently, to the coherence terms of the density matrix. More precisely, in order to gather quantitative evidence of the wake-field effect, we have investigated the relative populations of the fine structure components i) of the Balmer-α as the $3s_{1/2}$ / $3p_{1/2}$ and $3d_{3/2}$ / $3d_{5/2}$ s ratios in the case of krypton [13, 17, 19] and ii) of the Lyman-α as the $2p_{1/2}$ / $2p_{3/2}$ ratio in the case of argon [20].

In Figure 7, we present the case of H-like Kr ions initially populated by single excitation and transported through carbon foils. This collision system is very interesting in a sense that, in this case, the initial density matrix (the primary process) evaluated with Plane Wave Born Approximation calculations is well predicted. Hence, we test only the transport phase of our various theoretical approaches. With this figure, we better appreciate the sensitivity of the collisional effects, on one hand, and of the effect of the wake field, on the other hand. We note that the two quantum (MEA and QTT) approaches give similar results. For the $3p_{1/2}$ / $3s_{1/2}$ ratio, simulations with "wake off" (i.e., pure collisional approaches) predict first a decrease up to ~ 25 μg/cm² in clear contradiction with the observed experimental increase, which is, on the other hand, very well reproduced by the complete calculations that take into account the Stark coupling. This demonstrates that the wake field controls the evolution of hydrogen-like excited states. Moreover, the effective electric field deduced from the spatial extension of the wake field [7, 11] proves to be quite accurate. For the $3d_{3/2}$/$3d_{5/2}$ ratio that implies two states of the same $n\ell$ shell, collisional approaches predict, as expected, no evolution with the ion transit time. Taking into account the Stark mixing gives a good estimation of the asymptotic value. The ratio is slightly overestimated for intermediate thicknesses (~ 50-100 μg/cm²) but the disagreement is found to be around 10% at maximum.

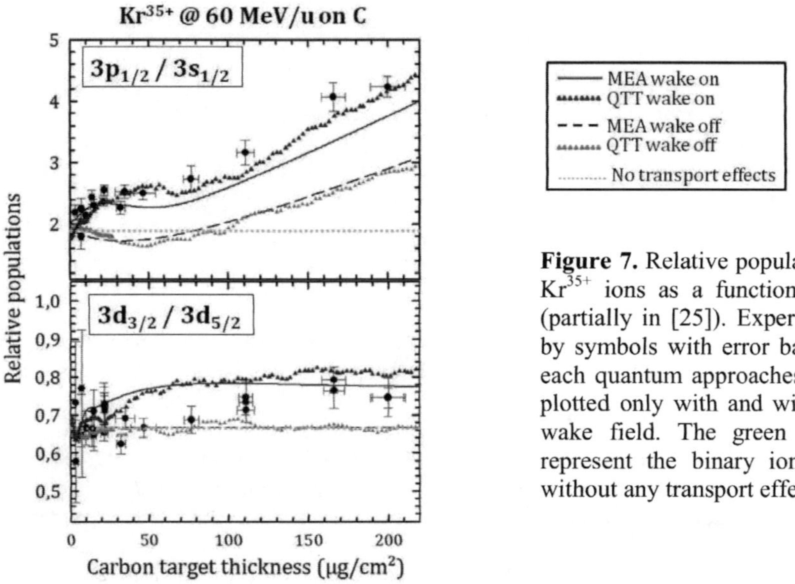

Figure 7. Relative population ratios in n = 3 for Kr^{35+} ions as a function of the foil thickness (partially in [25]). Experimental data are given by symbols with error bars. The predictions of each quantum approaches (MEA and QTT) are plotted only with and without the effect of the wake field. The green dashed straight lines represent the binary ion-atom collisions, i.e., without any transport effects.

For argon, the evolution of the $2p_{1/2}$ / $2p_{3/2}$ ratio is presented in Figure 8. As in the previous case, it varies significantly as a function of the target thickness. Here, when comparing the full predictions (black line) with those without Stark effect (dashed black line), we underline the major role of the wake field or, more precisely, the interplay between the wake field and the relativistic splitting of the energy level for this collision system. Changes as a function of ion penetration depth are well accounted for by our quantum transport simulation. Nevertheless, even if the general trend is satisfying, a clear discrepancy occurs at the "zero" thickness (i.e., the true single collision limit experimentally accessible with a gaseous target). For the primary process of charge transfer, the non relativistic calculations used here include only spin-independent interactions. As a result, the $n\ell j$

populations are linked to the nℓ populations through statistical factors leading to a ratio of 0.5. Experimentally, whatever the gas used (CH_4 or N_2), we found a value of 0.54 ± 0.01, ~ 10% larger than predicted indicating the presence of spin dependent relativistic corrections during the primary charge transfer even at a moderately relativistic speed. To estimate whether relativistic effects may explain the observed discrepancy, we have used the relativistic eikonal approximation to evaluate the ratio $2p_{1/2}$ / $2p_{3/2}$ [26]. A ratio of 0.52, at variance with 0.5, is predicted giving a hint that additional relativistic effects could be present. This ratio is nevertheless still insufficient to account for the experimental data. To have a better agreement for the thinner foils, a shift to a value of 0.56 is necessary (full blue line). Relativistic calculations of the one-electron capture density matrix at moderately relativistic velocities are missing to bring a final conclusion, the unprecedented accuracy of this type of measurements requiring thus new calculations.

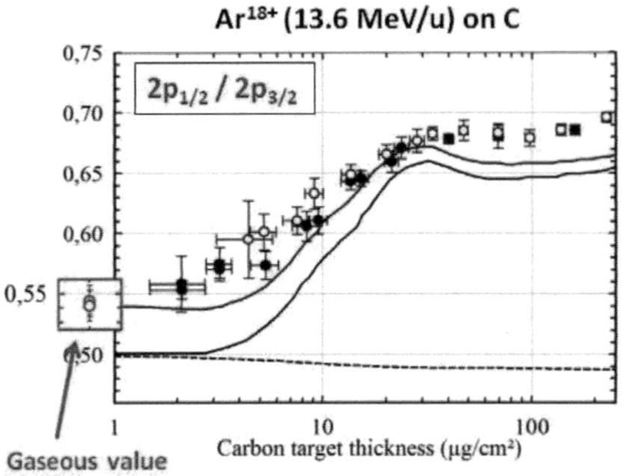

Figure 8. Lyman-α photon intensity ratio between $2p_{1/2}\rightarrow 1s$ and $2p_{3/2}\rightarrow 1s$ following transport of Ar^{17+} ions (initially Ar^{18+}) through carbon foils [26]. Symbols: experiment data. Note that the experimental value of the initial state (prior any transport) is shown (i.e., the gaseous value). Full black line: results predicted by QTT taking into account all the transport effects. Full blue line: as previous but with an adjusted ratio $R = \sigma_{2p1/2}/\sigma_{2p3/2}$. Dashed black line: predicted result when neglecting the wake field (equivalent to neglecting coherences).

5. Conclusions

The use of X-ray spectroscopy has allowed probing the ultra-fast dynamics of the ion-solid interaction as the X-ray emission is directly linked to the projectile ion excited state populations. The study of the evolution of high-ℓ Rydberg states as well as core nℓ and nℓ_j levels with the ion transit time inside the solid gives access to the response of the target electrons to the passage of the ion. The choice of collision partners (i.e., the projectile ion and the target whose thickness is varied) and of the collision velocity has allowed us to shed light on both features of the ion-solid interaction: the *collisional aspect* due to binary ion-atom processes (solid seen as an assembly of atoms) and the *collective response* via the wake field induced by the projectile. The theoretical description of the evolution of a small open system (the hydrogen-like projectile ion) in contact with a large reservoir (the solid) has been a great challenge. Whatever the approach used (Monte Carlo calculations or master equations), we succeeded to treat on the same footing all the competing processes occurring during the ion transport such as Stark mixing induced by the wake field following the projectile ion, elastic and inelastic collisions as well as radiative decay. The very good agreement between our experimental results and the simulations allow concluding that we have obtained a successful description of the ion transport in solid in the high velocity regime. In fact, it is remarkable to realize that our theoretical approaches first developed for dissipative processes in quantum optics implying a two- or three-level atom in interaction with an electromagnetic field may be generalized to a multi-level system in interaction with a reservoir including not only electromagnetic field but also collisions. In particular, we demonstrated that in ion-solid interaction, a collisional picture can be used to describe the high-ℓ Rydberg state

Journal of Physics: Conference Series **629** (2015) 012002 doi:10.1088/1742-6596/629/1/012002

populations and the major part of $n\ell$ core state populations. For these observables, it is worthwhile to mention that a classical transport gives similar results to those obtained with a quantum approach provided the radiative decay during the transport is negligible. Regarding in more details the $n\ell$ states and above all the relative $n\ell_j$ sub-state populations, we have shown that a pure collisional approach completely fails to reproduce their evolution inside the solid. Actually, the evolution of the relative fine-structure populations with the target thickness is well predicted provided Stark couplings induced by the wake field is accounted for. In more details, when the initial conditions are well known as, for instance, in the case of Kr^{35+} at 47 a.u. populated by single excitation in carbon, we found that the value of the wake field extracted from ion stopping power measurements is also correct to reproduce our experimental results. In other words, our experiment can be considered as the first direct measurement of the wake field. To go beyond, the sensitivity of our observable is such that we can also test the initial conditions (i.e., the initial density matrix) as seen in the case of Ar^{17+} at 23 a.u. populated by capture in carbon. For this collision system, a complete calculation on a $n\ell_j$ basis including relativistic effects is still needed to evaluate the one-electron capture density matrix proving, in particular, that the initial coherences in the capture process even at moderately relativistic velocities are not well theoretically described.

References

[1] Bethe H 1930 *Ann. Phys.* **5** 325
[2] Olivera G, Martinez A, Rivarola R and Fainstein P 1994 *Phys. Rev.* A **49** 603
[3] Schiwietz G, Wille U, Díez Muiño R, Fainstein R and Grande P 1996 *J. Phys.* B **29** 307
[4] Bohr N 1948 *K. Dansk. Vidensk. Selsk. Mat. Fys. Medd.* **18** 1
[5] Neufeld J and Ritchie R 1955 *Phys. Rev.* **98** 1632
[6] Echenique P, Flores F and Ritchie R 1990 Solid State Physics **43**, New York Academic 229: Ehrenreich H and Turnbull D
[7] Fuhr J, Ponce V, Garcia de Abajo F and Echenique P 1998 *Phys. Rev.* B **57** 9329
[8] Lamour E, Gervais B, Rozet J-P and Vernhet D 2006 *Phys. Rev.* A **73** 042715
[9] Seliger M, Reinhold C, Minami T, Schultz D, Pindzola M, Yoshida S, Burgdörfer J, Lamour E, Rozet J-P and Vernhet D 2007 *Phys. Rev.* A **75** 032714
[10] Minami T, Reinhold C and Burgdörfer J 2003 *Phys. Rev.* A **67** 022902
[11] Rozet J-P, Vernhet D, Bailly-Despiney I, Fourment C and Dubé L 1999 *J. Phys.* B **32** 4677
[12] Fourment C 2000 PhD thesis Université Pierre et Marie Curie
[13] Vernhet D, Rozet J-P, Bailly-Despiney I, Stephan C, Cassimi A, Grandin J-P and Dubé L 1998 *J. Phys.* B **31** 117
[14] Lamour E, Prigent C, Eberhardt B, Rozet J-P and Vernhet D 2009 *Rev. Sci. Instrum.* **80** 023103
[15] Adoui L 1995 PhD thesis Université Pierre et Marie Curie
[16] Cohen-Tanoudji C, Dupont-Roc J and Grynberg G 1988 *Processus d'interaction entre photons et atomes* Inter Editions/Editions du CNRS
[17] Reinhold C, Seliger M, Minami T, Yoshida S, Burgdörfer J, Mestayer J, Zhao W, Lancaster J and Dunning F 2007 *J. Phys. Conf. Series* **88** 012030
[18] Seliger M 2005 PhD thesis Technishe Universität Wien
[19] Lindblad G 1976 *Com. Math. Phys.* **48** 119
[20] Dalibard J, Castin Y and Molmer K 1992 *Phys. Rev. Lett* **68** 580
[21] Gervais B, Reinhold C and Burgdörfer J 1996 *Phys. Rev.* A **53** 3189
[22] Lamour E 1997 PhD thesis Université de Caen Basse Bormandie
[23] Reinhold C, Arbo D, Burgdörfer J, Gervais B, Lamour E, Vernhet D and Rozet J-P 2000 *J. Phys.* B **33** L111
[24] Vernhet D, Fourment C, Lamour E, Rozet J-P, Gervais B, Dubé L, Martin F, Minami T, Reinhold C, Seliger M and Burgdörfer J 2001 *Phys. Scr.* **T92** 233
[25] Reinhold C, Seliger M, Minami T, Schultz D, Burgdörfer J, Lamour E, Rozet J-P and Vernhet D 2007 *Nucl. Instrum. Meth.* B **261** 125
[26] Seliger M, Reinhold C, Minami T, Schultz D, Yoshida S, Burgdörfer J, Lamour E, Rozet J-P and Vernhet D 2008 *Phys. Rev.* A **77** 042713

Journal of Physics: Conference Series **629** (2015) 012003 doi:10.1088/1742-6596/629/1/012003

Ions interacting with complex molecular systems: The effect of a surrounding environment

Henning Zettergren[1]

[1] Department of Physics, Stockholm University, S-106 91 Stockholm, Sweden

E-mail: henning@fysik.su.se

Abstract. This paper highlight results from studies of keV-ion impact on complex molecules and molecular clusters, which have been carried out at the ARIBE facility in Caen (France) during the last decade. Studies of fullerenes, Polycyclic Aromatic Hydrocarbons (PAHs), and biomolecules are reviewed with focus on the effect of a surrounding environment when ions interact with weakly bound clusters of theses species. One common result is that charge and energy are rapidly shared between the individual molecules in the clusters, in contrast to e.g. weakly bound atomic clusters where the charge stay localized to a few atoms from which the electrons are removed during the collisions. Another important finding is that ion collisions may induce reactions within clusters such as e.g. proton transfer and different types of molecular growth processes. In the latter case, these processes may be driven by prompt non-statistical atom knockouts in billiard-ball like atom-atom collisions favouring highly reactive fragments. In contrast, statistical fragmentation in general yields different and less reactive fragments.

1. Introduction

During the last decades, there has been a strong development of different methods to produce pure macroscopic quantities of complex molecular systems and techniques to bring fragile systems such as e.g. biomolecules into the gas phase. Complex molecular systems have thus been extensively studied in the laboratory by means of various excitation and ionization tools such as e.g. photons, electrons, and energetic atoms/ions. When isolated molecules are excited in these interactions, the excess energy is typically redistributed across all internal degrees of freedom. This leads to statistical decay through electron emission, photon emission, or fragmentation processes on timescales exceeding the typical vibrational timescales of picoseconds. In such cases, the nature of excitation is unimportant - it is only the amount of excess energy which determines the fate of the molecules.

Studies of weakly bound clusters have also been subjects of strong research efforts. Such studies provide a bridge across the disciplines of physics and chemistry and aim to answer fundamental questions like: Do charge and energy stay localized to a few constituents in clusters which have been exposed to ionizing radiation? If not, how fast is the charge and energy communication? Is the nature of excitation important? Are the individual molecules protected from damage in the cluster environment? Are new fragmentation pathways observed in the cluster environment, which are not present for the isolated molecules? Is it possible to induce intracluster reactions and in that case what drives such processes?

In this paper, the first ever studies of collisions between keV-ions and weakly bound clusters of fullerenes, Polycyclic Aromatic Hydrocarbons (PAHs), and biomolecules are reviewed. These

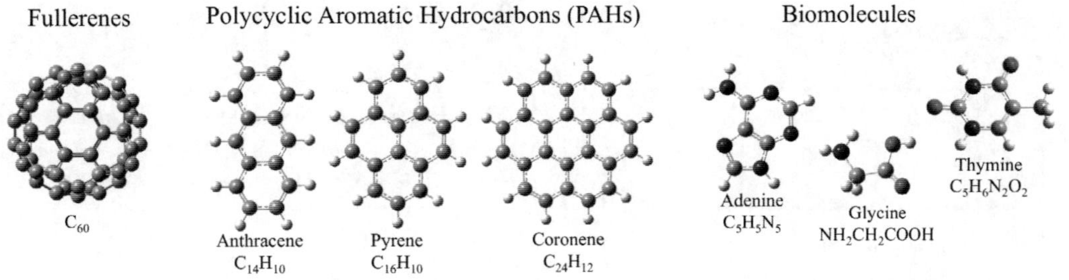

Figure 1. Examples of complex molecules which have been studied at the ARIBE facility.

pioneering experiments were carried out at the ARIBE facility in Caen (France) and have provided new insights into the above-mentioned fundamental questions. A few examples of the molecular structures used in these studies are shown in figure 1. Fullerenes are closed-cage all carbon structures where C_{60} is the most prominent family member. Due to its highly symmetric structure where all atoms experience identical chemical environments, C_{60} molecules serve as excellent model systems for complex molecular systems. Weakly bound clusters of fullerenes may be viewed as small pieces of fullerite - the solid state manifestation of C_{60}. The bonds between neighbouring C_{60} are of van der Waals type with C_{60}-C_{60} binding energies of only about 0.3 eV, which means that they may rotate freely and thus to a large extent keep their individual properties in fullerite and in clusters. Individual fullerenes (C_{60} and C_{70}) have been identified in space due to their characteristic IR-emission features [1], and possible signatures of fullerite (or clusters) have also been reported recently [2]. There they may coexist with Polycyclic Aromatic Hydrocarbons (PAHs) [3], which typically consist of fused benzene rings (cf. figure 1). PAHs are believed to be ubiquitous in space [4] and are thought to be the building blocks of interstellar dust grains. These findings raise questions about how fullerenes and PAHs are formed, react, respond to energy transfers, and the role of the surrounding (cluster) environment in such processes. One part of the motivation for the studies presented in this review was to shed light on these intriguing issues, which then may contribute to a better understanding of aspects of the origin and evolution of complex molecules in space [5].

Biomolecular precursors have also been identified in space and, consequently, the possibilities that the building blocks of life came to Earth from e.g. meteorite impacts have been much debated. This illustrates that it is important to understand how stable they are, how they are destroyed and how their fragments may later react to form new molecules. In the present case, keV-ions are used as tools to investigate the inherent properties of biomolecules such as e.g. DNA/RNA (nucleobases) and peptide/protein (amino acids) building blocks (cf. figure 1). Such studies also aim to further our knowledge on radiation damage processes at the single molecular level and to systematically investigate the effects of embedding a molecule inside a cluster. The latter is crucial to mimic a more realistic natural environment as for instance in an aqueous solution.

The paper is organized as follows. In section 2, the experimental multi-coincidence mass spectrometry techniques are briefly described. Sections 3 and 4 are devoted to discussions of results for collisions with monomer and cluster targets, respectively. The results are summarized in section 5 and an outlook with possible future experiments and challenges is also given there.

2. Experimental techniques

The experiments were performed at the ARIBE facility at GANIL in Caen, France. The experimental procedure has been discussed in detail elsewhere [6] and only a brief description is

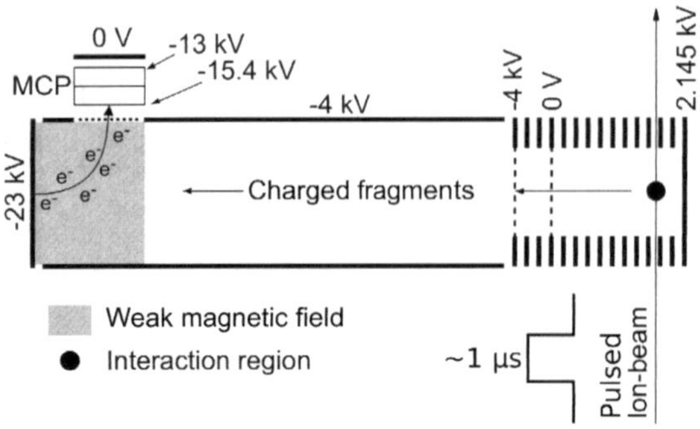

Figure 2. Schematic of the time-of-flight mass spectrometer used to analyze the positively charged products from collisions between keV-ions and complex molecular systems.

given here. Ion beams from an Electron Cyclotron Resonance (ECR) ion source are accelerated to keV energies. The ion beams are then pulsed at a repetition rate of a few kHz into microsecond long beam pulses, and interact with a broad distribution [7] of neutral weakly bound molecular clusters from a liquid nitrogen cooled cluster aggregation source. The positively charged collision products are analyzed with the aid of a linear time-of-flight mass spectrometer (cf. figure 2) [8] promptly after the beam pulses have left the interaction region. The collision products hit a gold coated steel plate at the end of the spectrometer, where secondary electrons emitted from the plate are guided to a microchannel (MCP) detector by a weak magnetic field. This arrangement give high detection efficiency, which is crucial for coincidence measurements of charged fragment from the same collision events.

3. Collisions with monomer targets

In the present collision energy regime, the projectiles are slower than the typical target valence electrons and electron capture is the dominant target ionization process. By varying the projectile charge, velocity, and mass, it is then possible to systematically investigate the effects of charging and heating molecular systems. In the present Section we summarize results for collisions with monomer targets - i.e. for targets in which the individual molecules are isolated from each other in vacuum.

3.1. High charge state projectiles

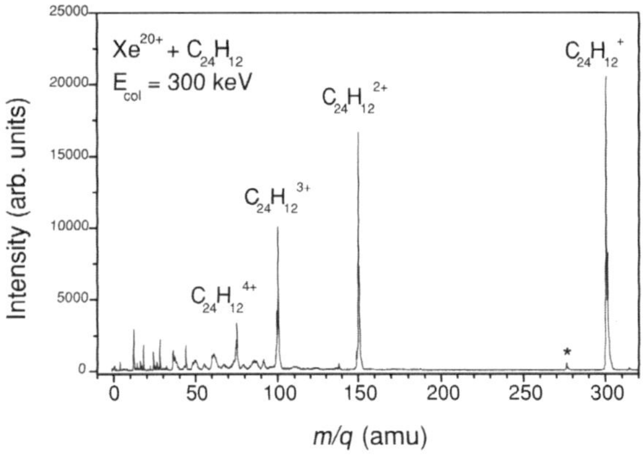

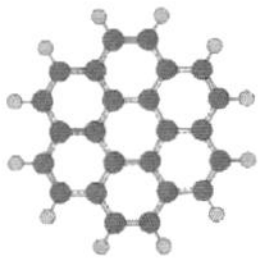

Figure 3. Left: Mass spectrum recorded for 300 keV Xe^{20+} + $C_{24}H_{12}$ collisions. (Reproduced with permission from [9] © 2011 APS) Right: Molecular structure of coronene ($C_{24}H_{12}$).

Collisions with highly charged projectiles are dominated by large impact parameter ionization

Journal of Physics: Conference Series **629** (2015) 012003 doi:10.1088/1742-6596/629/1/012003

through electron capture, where small amounts of energy is deposited directly in the collisions. Still, the molecules may be internally heated as the so formed multiply charged systems are born in their neutral geometry on the subfemtosecond timescale of the collision (vertical ionization), but later relax to minima on the potential energy surfaces from which they may statistically decay. For rigid molecular structures such as e.g. PAHs and fullerenes this heating processes is negligible and highly charged metastable systems may therefore survive on the experimental microsecond timescales. This is illustrated in figure 3, which shows the mass spectrum from collisions between 300 keV Xe^{20+} projectiles and coronene molecules ($C_{24}H_{12}$) [9]. The dominant peaks correspond to intact $C_{24}H_{12}^{q+}$ in charge states up to q=4, which are due to sequential electron capture processes at large impact parameters. A small fraction of the multiply charged systems (q=2-4) decay through statistical fragmentation processes, mainly through C_2H_2 and H-loss - the lowest dissociation energy channels at about 5 eV [10]. Thus, the relative intact peak intensities (including H-loss channels) reflect the sequence of ionization energies, i.e. the relative ionization cross sections. These are in good agreement with the results from a novel classical over-the-barrier model where PAHs are modeled as infinitely thin metal discs [11]. In this model, the energy barriers experienced by the active electrons are calculated along straight line ion trajectories. When these barriers become lower than the Stark shifted ionization energies the so called over-the-barrier criterion is fulfilled, and electrons are captured by the projectiles. The favorable comparison with experimental results show that the model catches the essentials of the collision dependent target-polarization effects, and that PAHs may be successfully described as small metallic objects when perturbed by an external electrostatic field [11]. Interestingly, fullerene molecules also display metallic behaviors in such interactions, as was recently demonstrated through comparisons between the electrostatic interaction energy for a metal sphere in the presence of a point charge and the corresponding results from molecular structure calculations [12, 13].

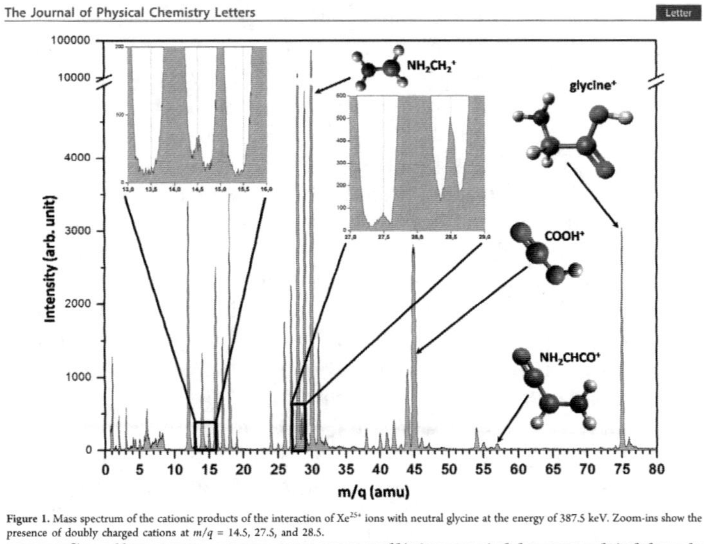

Figure 4. Mass spectrum recorded for collisions between 387.5 keV Xe^{25+} projectiles and glycine (NH_2CH_2COOH). (Reproduced with permission from [14] © 2013 ACS). The molecular structures of the singly charged parent ion and a few typical fragment ions are shown in the inset.

Smaller impact parameter collisions yield more highly charged coronene molecules (q>4), which promptly Coulomb explode into the small hydrocarbons with m/q<70 as can be seen in figure 3. This is commonly referred to as charge driven fragmentation. For smaller and less rigid systems such as e.g. biomolecules, internal heating through relaxation processes (cf. above) and Coulomb explosions become important already for moderately charged systems. This gives rise to a much richer fragmentation mass spectrum as illustrated by the example in figure 4 for collisions between 387.5 keV Xe^{25+} projectiles and the amino acid glycine (m/q=75) [14].

Journal of Physics: Conference Series **629** (2015) 012003 doi:10.1088/1742-6596/629/1/012003

Two of the most intense peaks in this spectrum have m/q=30 and m/q=45, which correspond to $NH_2CH_2^+$ and $COOH^+$, respectively. From the coincidence analysis it becomes evident that these fragments stem from the doubly charged parent molecule, i.e. from Coulomb exploding doubly charged glycine ($NH_2CH_2COOH^{2+}$). This is consistent with results from molecular structure calculations and ab-initio Molecular Dynamics simulations, which give information about the fragmentation dynamics leading to the formation of the doubly charged fragments (cf. the peaks at mass-to-charge ratios 14.5, 27.5 and 28.5 amu in the insets in figure 4). Interestingly, these calculations show that Coulomb explosions compete with neutral fragment emission following ultrafast intramolecular hydrogen migrations ($\sim$ 30 fs). Such ultrafast hydrogen transfers are expected to be important for other biomolecular systems [14]. This example and the recent study of γ-Aminobutyric Acid (NH_2-$(CH_2)_3COOH$) at ARIBE [15] clearly illustrate the synergy effects and power of combining coincidence time-of-flight mass spectrometry with ab-initio calculations and simulations.

3.2. Low charge state projectiles

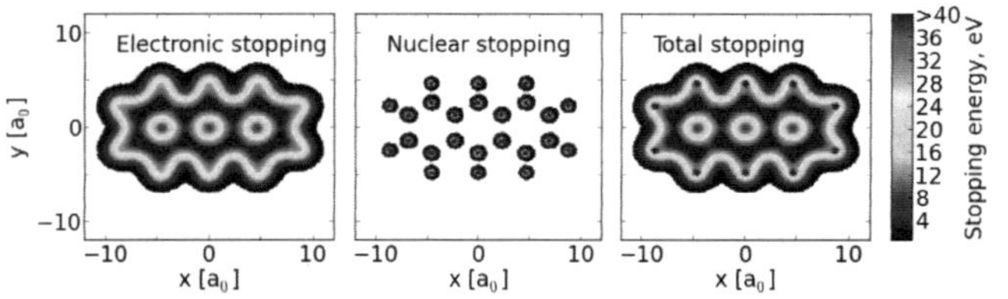

Figure 5. Model electronic (left panel), nuclear (middle panel), and total stopping energy (right panel) for 11.25 keV He + $C_{14}H_{10}$ collisions (cf. text).

For low charge state projectiles electrons are captured at small impact parameters, typically accompanied with substantial amounts of energy transfer to the molecule. Here the energy transfers may be localized to regions close to the projectile ion trajectories and are due to interactions with the electron clouds (electronic stopping) and with the individual atomic nuclei in the molecule (nuclear stopping). For light projectiles such as e.g. He^+, the excess energy is mainly due to electronic stopping processes, as the results from model calculations illustrate in figure 5 [16]. The left panel shows the pure electronic stopping for face on 11.25 keV He collisions with anthracene. This was calculated using the friction coefficients from Puska and Nieminen[17] for He traversing a free electron gas, where the molecular valance electron densities were taken from Density Functional Theory calculations. The corresponding nuclear stopping energies for binary He+C/H interactions using a screened Bohr potential is shown in the middle panel of figure 5, while the total (electronic+nuclear) stopping is shown in the right panel. The total stopping is clearly dominated by electronic stopping with energies on the order of 40 eV, which is well above the activation energies of about 5 eV for fragmentation [10].

After this subfemtosecond excitation process, the stopping energy is typically converted into internal heating followed by statistical decay processes [19]. This is for instance seen in the mass spectrum recorded in $He^+ + C_{14}H_{10}$ collisions [18], which is shown in the upper panel of figure 6. The most prominent peaks corresponds to intact singly and doubly charged anthracene. On the left hand side of these there are peaks due to fragmentation of moderately heated systems, yielding fragments corresponding to losses of H atoms and loss of C_2H_x (x=2-4). High energy deposits lead to multifragmentation processes as manifested by a broad distribution of $C_nH_x^+$ fragments with n=1-11. In this case the total fragmentation yield is (63±1)% , i.e. significantly larger than the (39±1)% recorded using 360 keV Xe^{20+} projectiles (cf. lower panel of figure 6). This clearly demonstrates that the target molecules on the average are much hotter when exposed to projectiles in low charge states compared to when they are exposed to high charge

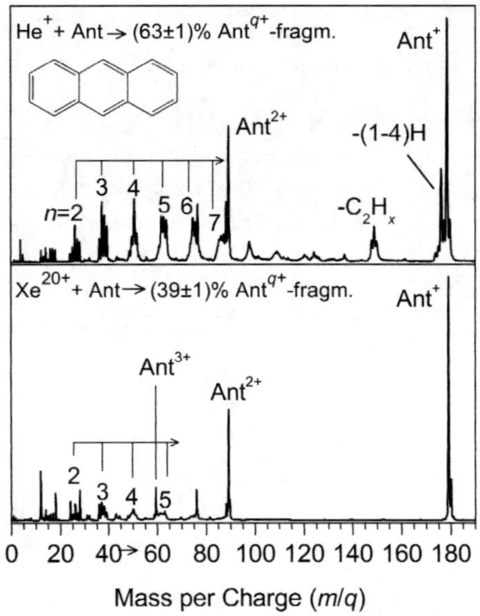

Figure 6. Mass-to-charge spectra for 11.25 keV He^+ + $C_{14}H_{10}$ (upper panel) and 360 keV Xe^{20+} + $C_{14}H_{10}$ collisions (lower panel). (Reproduced with permission from [18] © 2010 APS). The anthracene molecular structure is shown in the inset. The fragment mass distribution in the upper panel is typical for thermally activated PAHs in low charge states, leading to extensive fragmentation through statistical single- and multistep processes. The corresponding distribution in the lower panel is typical for charge driven PAH fragmentation, where multiply charged PAHs Coulomb explode into several small charged fragments (cf. text).

state projectiles. In the latter case, the molecules are typically multiply charged (cf. above) leading to several small singly charged fragments in Coulomb explosions, which explains why the fragment mass distribution is shifted towards smaller fragments in the lower panel of figure 6 and in figure 3.

4. Collisions with cluster targets

As mentioned in the Introduction, it is possible to systematically investigate the role of a surrounding environment by embedding the molecules inside a cluster. Here, key findings from such studies are briefly summarized. The first part is devoted to charge and energy flow processes when clusters are exposed to keV-ions, while the second part deals with intracluster reactions induced in such interactions.

4.1. Charge and energy flow processes

In 2003, the first study of keV-ions interacting with weakly bound clusters of complex molecules was carried out at the ARIBE facility in Caen [20]. The total mass spectrum from this study is shown in figure7, which is due to 400 keV Xe^{20+} projectiles colliding with a broad distribution of neutral clusters of fullerenes (including monomers). In this spectrum, intact singly and multiply charged fullerenes are the most dominant products. A large fraction of these are due to collisions with neutral monomers in the target, as they appear in the single-stop spectrum (the spectrum for events where only one charged ion is detected) and has a charge-state intensity distribution (C_{60}^{+}, C_{60}^{2+}, C_{60}^{3+} etc.) similar to those for a pure monomer target (cf. Ref. [20] for details). However, a substantial fraction of the singly charged intact monomers were shown to be detected in coincidence with each other or with smaller singly charged fragments. This means that they stem from fragmentation of multiply charged clusters, where the charges have been rapidly redistributed before the decay. Thus, in this study it was demonstrated that weakly bound clusters of fullerenes become highly conducting when charged, while being insulators as neutrals (i.e. like small pieces of bulk fullerite). This behavior is in strong contrast to weakly bound atomic Ar_n clusters where the charge stays localized to a few atoms when ionized by highly

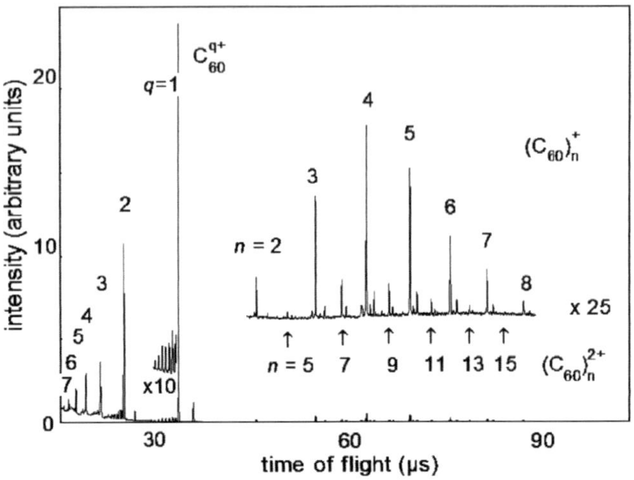

Figure 7. Time-of-flight mass spectrum of fullerenes and fullerene clusters, which have been ionized in collisions with Xe^{20+} projectiles at 400 keV. (Reproduced with permission from [20] © 2003 APS). The singly and multiply charged cluster distributions are shown in the inset, where the arrows indicate positions of odd numbered doubly charged clusters.

charged keV-projectiles [21]. To the right hand side of the singly charged monomer peak, there are distributions of intact singly and multiply charged clusters surviving on the experimental microsecond timescales, where the smallest cluster sizes for a given charge state (appearance sizes) were shown to be as small as n=5, 10, 21, and 33 for q=2, 3, 4, and 5, respectively.

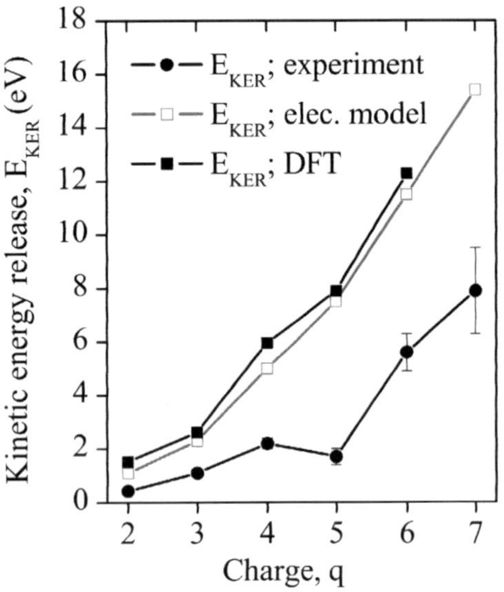

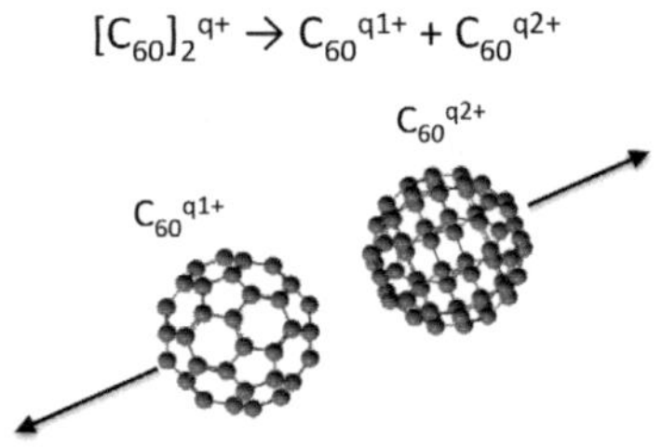

Figure 8. The kinetic energy releases in the fragmentation of $[C_{60}]_2^{q+} \rightarrow C_{60}^{q1+} + C_{60}^{q2+}$ as functions of charge, q1=q2 (even q), q1=q2+1 (odd q). (Reproduced with permission from [22] © 2009 AIP). The DFT results (solid squares) [22], the experimental results (solid circles) [23, 24], and the results from an electrostatic model (open squares) [23, 25].

As the charge is rapidly redistributed, the detection of two charged intact fullerene monomers in coincidence can be taken as a fingerprint for Coulomb explosion of a multiply charged dimer system. In Refs. [23] and [24] this was used to extract the relative ionization yields (by counting the number of such events) and the kinetic energy releases in the fragmentation processes (from the peak widths) as functions of the dimer charge state. Interestingly, the relative ionization yields display strong even-odd effects, which can only be explained if the charge communication

Journal of Physics: Conference Series **629** (2015) 012003 doi:10.1088/1742-6596/629/1/012003

takes place during the collisions, i.e. on the sub-femtosecond timescales (cf. Ref. [23] for details). This was later supported by combined Density Functional Theory (DFT) molecular structure calculations and modeling work [22]. However, the kinetic energy releases predicted from theory were shown to be significantly smaller than the measured ones (cf. figure 8). The reason for this discrepancy was attributed to internal heating by Coulomb explosions, i.e. that a substantial fraction (about 50%) of the potential energy is converted into internal energy of the separating monomers. The theoretical values are for internally cold systems and thus serve as upper limits for the kinetic energy release values.

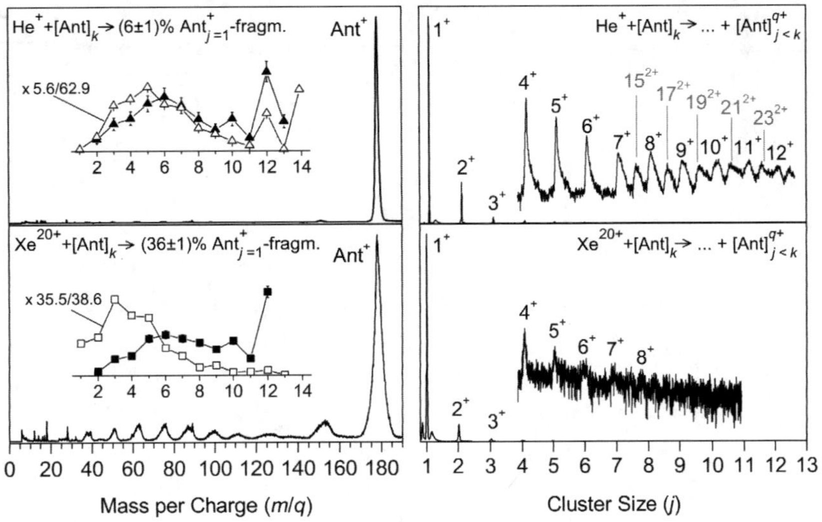

Figure 9. Left panels: Mass-to-charge spectra below m/q=190 for 11.25 keV He^+ + $[C_{14}H_{10}]_k$ (upper panel) and 360 keV Xe^{20+} + $[C_{14}H_{10}]_k$ collisions (lower panel). (Reproduced with permission from [18] © 2011 APS). Insets show relative $C_nH_x^+$ intensity distributions as functions of n for the cluster (filled symbols) and monomer target (open symbols) scaled by the ratio between the cluster and monomer fragmentation efficiencies. Right panels: Size-to-charge spectra for fragment clusters, $[C_{14}H_{10}]_{j<k}^{q+}$ produced in He^+/Xe^{20+} + $[C_{14}H_{10}]_k$ collisions. Insets: zoom-ins for j>4 with logarithmic intensity scales.

Interestingly, rapid charge and energy redistribution appear to be a general feature in collisions with weakly bound molecular systems, as it has also been demonstrated for biomolecular clusters [26, 27, 15, 28] as well as for PAH clusters [18, 29]. The latter is illustrated in figure9, which shows the mass spectra recorded in collisions between He^+/Xe^{20+} projectiles and anthracene ($C_{14}H_{10}$) clusters [18]. The monomer- and cluster-fragment regions of the mass spectra are shown in the left and right panels, respectively. The cluster distributions are rather similar for both projectiles, with a rapidly decreasing trend as a function of size, but the collision scenarios are different. In the He^+ case (upper panel), energy is mainly deposited in electronic stopping processes along the ion trajectories (cf. Sec. 3.2). This energy is redistributed and the clusters cool down by long sequences of neutral emissions of intact anthracene molecules, in most cases all the way down to the singly charged anthracene. Indeed, in this case monomer fragments stemming from clusters are significantly colder than when the anthracene monomer target is ionized by the same projectile at the same velocity (cf. figure 6). The fragmentation yield is about a factor of ten smaller, only (6±1)% in the cluster case. Thus, the monomers are protected from damage by the cluster environment. In the Xe^{20+} case (lower panel), only small amounts of energy is deposited directly in collisions (cf. Sec. 3.1). Surprisingly, the monomers are not protected from damage in collisions with Xe^{20+} as the fragmentation yield is similar to that for the pure monomer target (cf. figure 6). The reason for this is most likely heating by Coulomb explosions of multiply charged clusters, which was shown to be important for multiply charged fullerene dimers (cf. Fig 8).

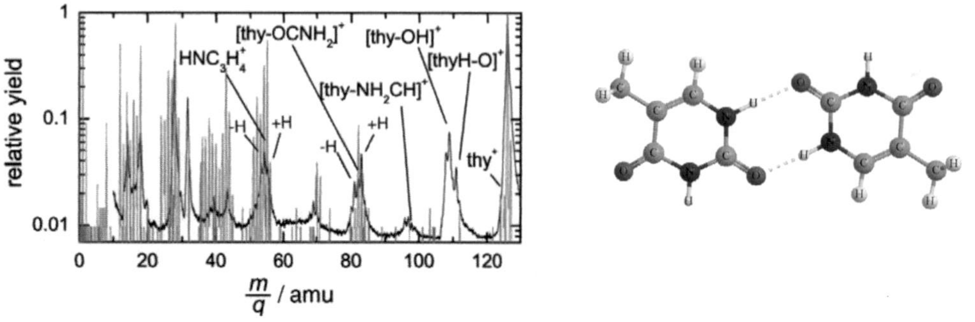

Figure 10. Left: Zoom-in of the mass spectrum recorded in collisions between 50 keV O^{5+}-projectiles and thymine clusters. The corresponding data for the monomer target is shown in grey. Right: Thymine dimer structure. (Reproduced with permission from [26] © 2006 John Wiley & Sons)

Another intriguing result is that the cluster environment may open up decay pathways which are closed in interactions with the isolated molecules. This was demonstrated in collisions between 60 keV C^{5+} and nucleobase monomers and clusters [26]. Figure 10 shows that the mass spectra for collisions with thymine targets display similar features, but there are important differences. The most striking effect is the appearance of new channels in the cluster case corresponding to O- and OH-loss. Using coincidence data, it was shown that this channel appears already in collisions with the smallest clusters (dimers) [26]. The most likely explanation for this new channel is that the intermolecular O-H bonds weakens the intramolecular bonds in the clusters, which favor the loss of a single O and a single H atom involved in the hydrogen bonds (cf. figure 10). This illustrates another type of effect due to the cluster environment which so far has not been observed with PAH and fullerene targets.

4.2. Intracluster reactions

It is well established that heated fullerenes emit C_2-molecules rather than single carbon atoms in statistical fragmentation processes [31]. This simply reflects the large differences in dissociation energies, about 10 eV [32] and 15 eV [33], respectively. However, despite intense research the actual fullerene *formation* mechanism is still debated, as it is not fully understood why C_{60} are formed so much more abundantly than other fullerenes in nature (e.g. in a fire flame). Possible explanations involve e.g. bottom-up mechanisms in which individual C-atoms or C_2-molecules are ingested by smaller fullerenes [34], top-down mechanisms from decay of hot giant fullerenes [35], or emission of electromagnetic radiation [36]. These rely on the lower reactivity [34], more stable structure [35], and more effective radiative cooling of C_{60} [36] with respect to neighboring fullerene sizes.

It has been demonstrated that larger fullerenes may be formed from smaller ones when fullerene films and fullerene clusters are exposed to various kind of photon fluxes [37, 38, 39]. This typically leads to broad distributions of fullerene sizes with an even number of carbon atoms,

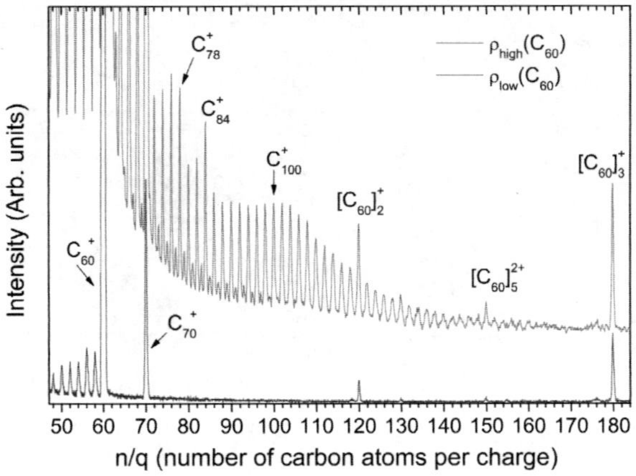

Figure 11. Zoom-in of the mass spectra due to 400 keV Xe^{20+}+ $[C_{60}]_n$ collisions. (Reproduced with permission from [30] © 2010 AIP). The spectra were recorded with low (ρ_{low}) and high (ρ_{high}) fullerene monomer densities in the cluster source, i.e. with small and large neutral clusters, respectively. Fullerene dimers, trimers, and pentamers are indicated with $[C_{60}]^+_2$, $[C_{60}]^+_3$, and $[C_{60}]^{2+}_5$ and fullerene monomers with C^+_n.

reflecting the higher stabilities of those with respect to the odd-numbered ones. Similar mass distributions were observed in collisions between Xe^{20+} projectiles and weakly bound clusters of fullerenes [30]. As shown in figure 11 these molecular fusion reactions are only seen with *large* cluster targets. The scenario here is that the individual molecules are destroyed along cluster-penetrating ion trajectories and that the surrounding environment (i.e. large clusters) allows the fragments to react with neighboring intact fullerenes or with other fragments before the clusters completely disintegrate. The measured kinetic energies of the fusion products in Ref. [30] suggest that both size-up and size-down processes are responsible for the mass broad distribution in figure 11. A large fraction of these fusion products corresponds to the most stable closed caged fullerenes, as concluded from the strong correlation between the measured relative peak intensities and the calculated relative stabilities as a function of fullerene size.

More recently, a novel type of molecular fusion reaction has been observed in collisions between He^{2+} projectiles and clusters of fullerenes [33]. This is illustrated in figure 12, where the left part shows the mass spectra for all collision events (upper panel) and events correlated with one or several intact C^+_{60} ions (lower panel). The total mass spectrum is rather similar to the one recorded in Xe^{20+} + $[C_{60}]_n$ collisions (cf. figure 7), but there is significantly more intensity in the dimer region. A large fraction of this intensity appears in the coincidence spectrum and is due to three peaks at the positions for 120, 119, and 118 carbon masses per atomic unit of charge. Surprisingly, the most prominent peak corresponds to an odd number of carbon atoms. These are produced in prompt, nonstatistical, knockouts in Rutherford-like binary He+C scattering processes. The so formed C^+_{59} ions rapidly react with neighboring C_{60} molecules in the clusters efficiently forming dumbbell shaped molecular fusion products while the clusters explode (cf. the right part of figure12). This scenario was supported by molecular dynamics simulations, which show that C^+_{59} are highly reactive and thus easily form thermodynamically stable dumbbell C^+_{119} systems [33].

Direct evidence for prompt knock-outs have recently been reported in PAH^+ + He collisions [40, 41, 42], at significantly lower center of mass collision energies (110 eV) where nuclear stopping is the dominant energy loss process. Thus, the molecules are only slightly heated internally (by about 4 eV) in electronic stopping processes and the fragments, which are as highly reactive as C^+_{59}, may therefore survive on the experimental microsecond timescales. This clearly illustrates that it is a general mechanism, and one would therefore expect to also see such non-statistical driven molecular fusion reactions for complex molecular cluster systems other than fullerenes. However, this has not yet been reported in the literature for collisions with pure PAH or mixed

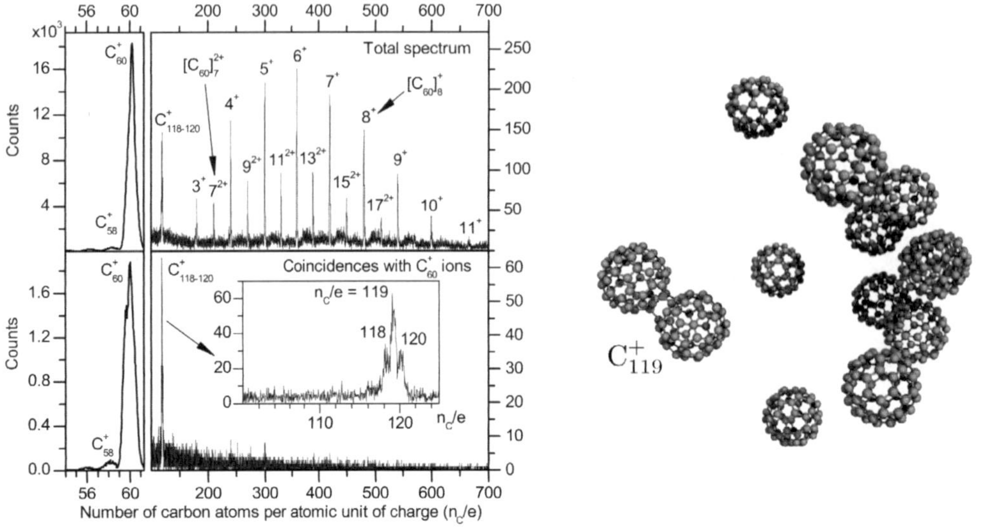

Figure 12. Left: The measured mass-to-charge distributions following 22.5 keV He^{2+}+ $[C_{60}]_n$ collisions. (Reproduced with permission from [33] © 2013 APS). The top and bottom panels show the distributions including all events (total spectrum) and for events correlated with one or several intact C_{60}^{+} ions, respectively. Note the different scales for the left and right panels. The inset in the lower right panel shows a zoom-in on the mass region from 100 to 125 carbon masses per atomic unit of charge. Right: Snapshot from molecular dynamics simulations where a covalently bound dumbbell C_{119}^{+} system has been formed in a C_{59}^{+}+C_{60} collision when a cluster of thirteen fullerenes explodes.

fullerene/PAH cluster targets. According to pilot molecular structure calculations and molecular dynamics simulations, the kinetic energy required to fuse defect PAH structures with intact PAHs/fullerenes is too high, but it should be possible with molecular systems which are less rigid and more easily bond to the defects [16]. This calls for future studies of mixed clusters.

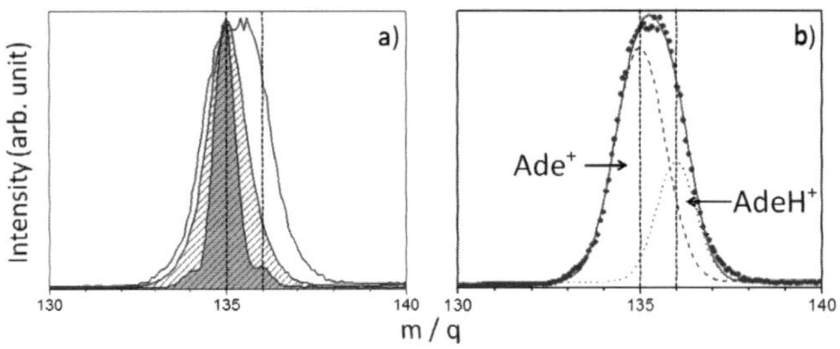

Figure 13. Left: The singly charged adenine monomer peaks recorded in collisions between 37.2 keV O^{3+} ions and adenine monomers (grey area), pure adenine clusters (hatched area), and hydrated adenine clusters (white area). Right: A fit to the monomer peak from collisions with hydrated clusters show contributions from singly charged (m/q=135) and protonated adenine (m/q=136). (Reproduced with permission from [28] © 2012 IOP)

Other types of intracluster reactions have been observed for biomolecules embedded in water clusters [28]. The left panel of figure 13 shows the singly charged adenine monomer peaks following collisions between 37.2 keV O^{3+} ions and adenine monomers, pure adenine clusters, and hydrated adenine clusters. The latter is significantly broader and centered around a higher mass to charge ratio compared to the other ones. This is due to formation of protonated adenine, as revealed from the fit shown in the right panel of figure 13. Here, the formation mechanism is most likely similar to that observed in 12 MeV/u Ni^{25+} impact on pure water clusters [43], where it was demonstrated that water may capture protons from nearby ionized water molecules. The proton affinity is larger for adenine than for water, which readily explains the efficient formation of protonated adenine inside the water cluster.

5. Summary and outlook

To summarize, the first ever studies of ion impact on various kinds of clusters, such as e.g. pure clusters of fullerenes, PAHs, nucleobases, amino acids and hydrated nucleobases, have been performed at the ARIBE facility in Caen, France. One common conclusion from these studies is that charge and energy are rapidly redistributed among the clusters constituents. In addition, different types of intracluster reactions may be induced depending on the actual collision conditions. The cluster environment plays an important role here, proton transfer reactions have for instance been observed for nucleobases in water but not for pure clusters of nucleobases. Intracluster reactions also depend on the size of the clusters and the projectile type, as e.g. certain fullerene fusion reactions are only seen when large clusters are exposed to Xe^{20+} projectiles, while He^{2+} ions ignite different fusion processes.

In the near future it will be possible to study collisions between keV-ions and *mass-selected* cluster ions with the aid of the novel PIBALE (Plateforme d'Irradiation de Biomolécules et d'Agrégats Libres et Environnés) setup at ARIBE. This will be a tremendous advantage when interpreting the experimental results, as the collision system will be well-defined in contrast to the previous studies of broad distributions of neutral cluster sizes. Another important aspect is to determine the energy being deposited in the interactions, which is an unknown parameter and hampers the interpretations of the results. This is truly challenging from both an experimental and a theoretical point of view. To realize this, common efforts will be essential, for instance within the COST Action XLIC and the European Associated Laboratory (DYNAMO) consisting of research groups from ARIBE, Stockholm University (Sweden), and Universidad Autónoma de Madrid (Spain).

In a somewhat longer perspective, the unique expertise and experience gained from studies of complex molecular systems at ARIBE could serve as an excellent platform for even more advanced experiments, possibly including electrostatic storage devices - traps and/or rings. There ions could be monitored and manipulated on longer timescales. New such facilities, including cryogenically cooled ones, are now being built and commissioned in e.g. Japan [44], Germany [45], and Sweden [46, 47]. The cooling option is a way of controlling the experimental conditions as the ions are in more well-defined quantum states than in room temperature environments. Having such a storage device would further broaden the range of unique studies at ARIBE, and drive the relevant technology and science forward in this new emerging community.

Acknowledgments

This work was supported by the Swedish Research Council (Contract No 621-2012-3660). The author would like to acknowledge the COST action CM1204 "XUV/X-ray light and fast ions for ultrafast chemistry (XLIC)" and the research groups in Caen, Madrid, and Stockholm.

References

[1] Cami J, Bernard-Salas J, Peeters E and Malek S E 2010 *Science* **329** 1180

[2] Evans A, van Loon J T, Woodward C E, Gehrz R D, Clayton G C, Helton L A, Rushton M T, Eyres S P S, Krautter J, Starrfield S and Wagner R M 2012 *Mon. Nat. R. Astron. Soc.* **421** L92

[3] Berne O and Tielens A G G M 2012 *Proc. Natl. Acad. Sci.* **109** 401

[4] Tielens A G G M 2008 *Annu. Rev. Astron. Astr.* **46** 289

[5] Tielens A G G M 2013 *Rev. Mod. Phys.* **85** 1021

[6] Bergen T, Biquard X, Brenac A, Chandezon F, Huber B A, Jalabert D, Lebius H, Maurel M, Monnand E, Opitz J, Pesnelle A, Pras B, Ristori C and Rocco J C 1999 *Rev. Sci. Instrum.* **70** 3244

[7] Maul J, Berg T, Marosits E, Schönhense G and Huber G 2006 *Phys. Rev.* B **74** 161406

[8] Chandezon F, Huber B A and Ristori C 1994 *Rev. Sci. Instrum.* **65** 3344

[9] Ławicki A, Holm A I S, Rousseau P, Capron M, Maisonny R, Maclot S, Seitz F, Johansson H A B, Rosén S, Schmidt H T, Zettergren H, Manil B, Adoui L, Cederquist H and Huber B A 2011 *Phys. Rev.* A **83** 022704

[10] Holm A I S, Johansson H A B, Cederquist H and Zettergren H 2011 *J. Chem. Phys.* **134** 044301

[11] Forsberg B O, Alexander J D, Chen T, Pettersson A T, Gatchell M, Cederquist H and Zettergren H 2013 *J. Chem. Phys.* **138** 054306

[12] Zettergren H, Forsberg B O and Cederquist H 2012 *Phys. Chem. Chem. Phys.* **14** 16360

[13] Zettergren H and Cederquist H 2012 *Phys. Chem. Chem. Phys.* **14** 16770

[14] Maclot S, Piekarski D G, Domaracka A, Méry A, Vizcaino V, Adoui L, Martín F, Alcamí M, Huber B A, Rousseau P and Díaz-Tendero S 2013 *J. Phys. Chem. Lett* **4** 3903

[15] Capron M, Diaz-Tendero S, Maclot S, Domaracka A, Lattouf E, Ławicki A, Maisonny R, Chesnel J Y, Méry A, Poully J C, Rangama J, Adoui L, Martín F, Alcamí M, Rousseau P and Huber B A 2012 *Chemi. Eur. J.* **18** 9321

[16] Chen T, Gatchell M, Stockett M H, Alexander J D, Zhang Y, Rousseau P, Domaracka A, Maclot S, Adoui L, Huber B A, Schlathölter T, Schmidt H T, Zettergren H and Cederquist H 2014 *J. Chem. Phys.* **140** 224306

[17] Puska M J and Nieminen R M 1983 *Phys. Rev.* B **27** 6121

[18] Holm A I S, Zettergren H, Johansson H A B, Seitz F, Rosén S, Schmidt H T, Ławicki A, Rangama J, Rousseau P, Capron M, Maisonny R, Adoui L, Méry A, Manil B, Huber B A and Cederquist H 2010 *Phys. Rev. Lett.* **105** 213401

[19] Martin S, Chen L, Brédy R, Montagne G, Ortega C, Schlathölter T, Reitsma G and Bernard J 2012 *Phys. Rev.* A **85** 052715

[20] Manil B, Maunoury L, Huber B A, Jensen J, Schmidt H T, Zettergren H, Cederquist H, Tomita S and Hvelplund P 2003 *Phys. Rev. Lett.* **91** 215504

[21] Tappe W, Flesch R, Rühl E, Hoekstra R and Schlathölter T 2002 *Phys. Rev. Lett.* **88** 143401

[22] Zettergren H, Wang Y, Lamsabhi A M, Alcamí M and Martín F 2009 *J. Chem. Phys.* **130** 224302

[23] Zettergren H, Schmidt H T, Reinhed P, Cederquist H, Jensen J, Hvelplund P, Tomita S, Manil B, Rangama J and Huber B A 2007 *Phys. Rev.* A **75** 051201

[24] Zettergren H, Schmidt H T, Reinhed P, Cederquist H, Jensen J, Hvelplund P, Tomita S, Manil B, Rangama J and Huber B A 2007 *J. Chem. Phys.* **126** 224303

[25] Zettergren H, Schmidt H T, Cederquist H, Jensen J, Tomita S, Hvelplund P, Lebius H and Huber B A 2002 *Phys. Rev.* A **66** 032710

[26] Schlathölter T, Alvarado F, Bari S, Lecointre A, Hoekstra R, Bernigaud V, Manil B, Rangama J and Huber B 2006 *ChemPhysChem* **7** 2339

[27] Maclot S, Capron M, Maisonny R, Ławicki A, Méry A, Rangama J, Chesnel J Y, Bari S, Hoekstra R, Schlathölter T, Manil B, Adoui L, Rousseau P and Huber B A 2011 *ChemPhysChem* **12** 930

[28] Domaracka A, Capron M, Maclot S, Chesnel J Y, Méry A, Poully J C, Rangama J, Adoui L, Rousseau P and Huber B A 2012 *J. Phys. Conf. Ser.* **373** 012005

[29] Johansson H A B, Zettergren H, Holm A I S, Seitz F, Schmidt H T, Rousseau P, Ławicki A, Capron M, Domaracka A, Lattouf E, Maclot S, Maisonny R, Manil B, Chesnel J Y, Adoui L, Huber B A and Cederquist H 2011 *Phys. Rev.* A **84** 043201

[30] Zettergren H, Johansson H A B, Schmidt H T, Jensen J, Hvelplund P, Tomita S, Wang Y, Martín F, Alcamí M, Manil B, Maunoury L, Huber B A and Cederquist H 2010 *J. Chem. Phys.* **133** 104301

[31] Campbell E E B and Rohmund F 2000 *Rep. Prog. Phys.* **63** 1061

[32] Tomita S, Andersen J U, Gottrup C, Hvelplund P and Pedersen U V 2001 *Phys. Rev. Lett.* **87** 073401

[33] Zettergren H, Rousseau P, Wang Y, Seitz F, Chen T, Gatchell M, Alexander J D, Stockett M H, Rangama J, Chesnel J Y, Capron M, Poully J C, Domaracka A, Méry A, Maclot S, Schmidt H T, Adoui L, Alcamí M, Tielens A G G M, Martín F, Huber B A and Cederquist H 2013 *Phys. Rev. Lett.* **110** 185501

[34] Dunk P W, Kaiser N K, Hendrickson C L, Quinn J P, Ewels C P, Nakanishi Y, Sasaki Y, Shinohara H, Marshall A G and Kroto H W 2012 *Nature Comms.* **3** 855

[35] Irle S, Zheng G and Morokuma K 2006 *J. Phys. Chem.* B **110** 14531
[36] Hansen K and Campbell E 1996 *J. Chem. Phys.* **104** 5012
[37] Yeretzian C, Hansen K, Diederich F and Whetten R L 1992 *Nature* **359** 44
[38] Mitzner R, Winter B, Kusch C, Campbell E and Hertel I 1996 *Z. Phys.* D **37** 89
[39] Hedén M, Hansen K and Campbell E E B 2005 *Phys. Rev.* A **71** 055201
[40] Gatchell M, Stockett M H, Rousseau P, Chen T, Kulyk K, Schmidt H T, Chesnel J Y, Domaracka A, Méry A, Maclot S, Adoui L, Støchkel K, Hvelplund P, Wang Y, Alcamí M, Huber B A, Martín F, Zettergren H and Cederquist H 2014 *Int. J. Mass Spectrom.* **365-366** 260
[41] Stockett M H, Zettergren H, Adoui L, Alexander J D, Berzins U, Chen T, Gatchell M, Haag N, Huber B A, Hvelplund P, Johansson A, Johansson H A B, Kulyk K, Rosen S, Rousseau P, Støchkel K, Schmidt H T and Cederquist H 2014 *Phys. Rev.* A **89** 032701
[42] Stockett M H, Gatchell M, Alexander J D, Berzins U, Chen T, Farid K, Johansson A, Kulyk K, Rousseau P, Støchkel K K, Adoui L, Hvelplund P, Huber B A, Schmidt H T, Zettergren H and Cederquist H 2014 *Phys. Chem. Chem. Phys.* **16** 21980
[43] Adoui L, Cassimi A, Gervais B, Grandin J P, Guillaume L, Maisonny R, Legendre S, Tarisien M, Lopez-Tarifa P, Politis M F, du Penhoat M A H, Vuilleumier R, Gaigeot M P, Tavernelli I, Alcamí M and Martín F 2009 *J. Phys.* B **42** 075101
[44] Nakano Y, Morimoto W, Majima T, Matsumoto J, Tanuma H, Shiromaru H and Azuma T 2012 *J. Phys. Conf. Ser.* **388** 142027
[45] Lange M, Froese M, Menk S, Varju J, Bastert R, Blaum K, Lopez-Urrutia J R C, Fellenberger F, Grieser M, von Hahn R, Heber O, Kuehnel K U, Laux F, Orlov D A, Rappaport M L, Repnow R, Schroeter C D, Schwalm D, Shornikov A, Sieber T, Toker Y, Ullrich J, Wolf A and Zajfman D 2010 *Rev. Sci. Instrum.* **81** 055105
[46] Thomas R D, Schmidt H T, Andler G, Björkhage M, Blom M, Brännholm L, Bäckström E, Danared H, Das S, Haag N, Hallden P, Hellberg F, Holm A I S, Johansson H A B, Källberg A, Kallersjö G, Larsson M, Leontein S, Liljeby L, Löfgren P, Malm B, Mannervik S, Masuda M, Misra D, Orban A, Paal A, Reinhed P, Rensfelt K G, Rosen S, Schmidt K, Seitz F, Simonsson A, Weimer J, Zettergren H and Cederquist H 2011 *Rev. Sci. Instrum.* **82** 065112
[47] Schmidt H T, Thomas R D, Gatchell M, Rosen S, Reinhed P, Löfgren P, Brännholm L, Blom M, Björkhage M, Bäckström E, Alexander J D, Leontein S, Hanstorp D, Zettergren H, Liljeby L, Källberg A, Simonsson A, Hellberg F, Mannervik S, Larsson M, Geppert W D, Rensfelt K G, Danared H, Paal A, Masuda M, Hallden P, Andler G, Stockett M H, Chen T, Kallersjö G, Weimer J, Hansen K, Hartman H and Cederquist H 2013 *Rev. Sci. Instrum.* **84** 055115

Journal of Physics: Conference Series **629** (2015) 012004 doi:10.1088/1742-6596/629/1/012004

Biological systems: from water radiolysis to carbon ion radiotherapy

Michael Beuve[1], **Jean-Michel Moreau**[2], **Claire Rodriguez**[3] **and Etienne Testa**[1]

[1] IPNL,UMR 5822, Université de Lyon, Université Lyon 1, CNRS/IN2P3, 4 rue E. Fermi 69622 Villeurbanne cedex, France
[2] LIRIS, Université de Lyon, Université Lyon 1,France
[3] LRCM, Université de Lyon, Université Lyon 1,France

E-mail: michael.beuve@univ-lyon1.fr

Abstract. Hadron therapy is an innovative cancer treatment method based on the acceleration of light ions at high energy. In addition to their interesting profile of dose deposition, which ensures accurate targeting of localized tumors, carbon ions offer biological properties that lead to an efficient treatment for radio- and chemo-resistant tumors and to provide a boost for tumors in hypoxia. This paper is a short review of the progress in theoretical, experimental, fundamental and applied research, aiming at understanding the origin of the biological benefits of light ions better. As a limit of such a vast and multidisciplinary domain, this review adopts the point of view of the physicists, leaning on results obtained in connection with CIMAP's IRRABAT platform.

1. Introduction

Interaction of fast ions with biological systems constitutes one aspect of the interdisciplinary researches performed with ion-beam facilities. This domain is as rich as it is complex since it encompasses several orders of magnitude in both space and time. The shortest space and time scale corresponds to atomic collisions, which may be as short as 10^{-18} s for the interaction of fast ions with individual atoms. At the opposite end of this domain, late effects – like cancer induction, chromosomal instability or organ dysfunctions – may appear or remain several years after irradiations. While irradiations may be limited to a very localized region, the whole behavior of an organ may be affected, possibly leading to human death, in particular when the irradiation dose and spatial extension are high. Between these two extreme scales, stands a great number of mechanisms, including for instance: the transport of the primary ejected electrons, the relaxation of the ionized and excited molecules, which may lead to direct damage in biological targets and to radical species and associated biochemical reactions. These early physical and chemical stages are followed by numerous and complex cell responses, such as the triggering of mechanisms to check DNA, to repair its damage, to manage the oxidative stress or to induce cell death. The numerous biological endpoints that have been studied reveal the complexity and the diversity of this biological response. These endpoints may involve particular structures of cells at the molecular scale (tracking of protein activities, damage in DNA, protein or lipid) or at the sub-cellular scale (chromosomes, nucleus, membranes, mitochondria…) and may concern cell organization (3D cell culture, tissues, organs, body). The domain of low doses

($\ll$Gy) and low dose rates (e.g. lower than cGy/hour) is mainly explored in order to estimate the risks of ionizing radiations for heath, while the domain of the so-called high doses (>Gy) and high-dose rates (Gy/minute), is often addressed in the context of cancer treatment by external radiotherapy. The title of the present paper therefore refers to an extremely large topic, actually too vast to be summarized in one paper. Our contribution aims at providing, as illustrations of this promising research field, some key examples of scientific progress, mainly leaning on results obtained in connection with CIMAP's IRRABAT platform or made within an IPNL-GANIL collaboration, focusing in particular on the effects of high Linear Energy transfer (LET) in the context of research for hadrontherapy with light ions. Moreover, this contribution is presented with a physicist's point of view in mind.

2. Hadrontherapy with carbon ions

2.1. A brief history of hadrontherapy

The history of hadrontherapy goes back to the beginning of radiotherapy, and therefore, to the discovery of X-rays by Wilhem Conrad Röntgen in 1895. It may sound quite amazing today that, just one year after this discovery, a first clinical trial of treatment by X-rays was attempted in France (Lyon) by Victor Despeignes. At the opposite, many more years separated the invention of the cyclotron, by Ernest Lawrence in 1930, and the proposition in 1946 by Robert Wilson [1] to treat patients with fast ions. The idea was later applied in the 1950's with a proton beam at the Lawrence Berkeley Laboratory (LBL) and followed by other experiments e.g. in 1957 at the University of Uppsala, in 1961 at the MGH (Massachusetts General Hospital). These led to a first hospital-based proton facility at the Loma Linda University Medical Center. The first treatment trials with fast heavy ions were undertaken at Berkeley in the 1970's, but the first clinical center dedicated to cancer treatment with carbon ions was built at Chiba, Japan, (HIMAC Center [NIRS]) in 1994. Presently, eight centers for carbon therapy have been built in countries like Japan (1994 Chiba (HIMAC) - 2001 Hyogo (HIBMC) - 2010 Gunma University (GHMC)), Germany (1997 Pilot project in Darmstadt (GSI) - 2009 Heidelberg (HIT) - Marburg (CPT)), Italy (2013 Pavia (CNAO)), Austria (2013 MedAustron (first patient in 2015)), China (Langzhou (IMP)). Eleven new centers are in project.

2.2. Physical properties of carbon beams

As for protons, the so-called inverted dose-depth profile with the maximum dose deposition at the end of the ion path (Bragg peak), is a key feature of carbon ions. With respect to proton beam, the lateral straggling is even lower, offering more accuracy. While the nuclear fragmentation of carbon ions leads to a larger dose beyond the position of the Bragg peak in comparison to protons, this dose is drastically lower than that deposited by the high-energy photons used in radiotherapy. Moreover, fragmentation offers many opportunities to control the irradiation of patients, on-line or off-line: it is probable that the detection of prompt γ-rays [2, 3], secondary protons [4, 5] or the 511 keV γ-rays produced by positron emitters should increase the safety of treatments [6] in the future.

2.3. Biological interest of carbon ions

However, the major interest of carbon ions resides in the specific biological effect induced by carbon ions with respect to protons or any form of conventional radiotherapy. This effect is twofold and characterized by two indicators, the *Relative Biological Efficiency* (RBE) and the *Oxygen Enhancement Ratio* (OER). These coefficients will be described in the next sections, focusing in particular on the cell survival endpoint, since this quantity is strongly related to the local control of tumors. The dose dependency of radiation-induced biological effects is a complex phenomenon. At low dose, some effects, like hypersensitivity and enhanced radioresistance, have been observed and are still not clearly understood. When plotted in logarithmic scale, cell

Journal of Physics: Conference Series **629** (2015) 012004 doi:10.1088/1742-6596/629/1/012004

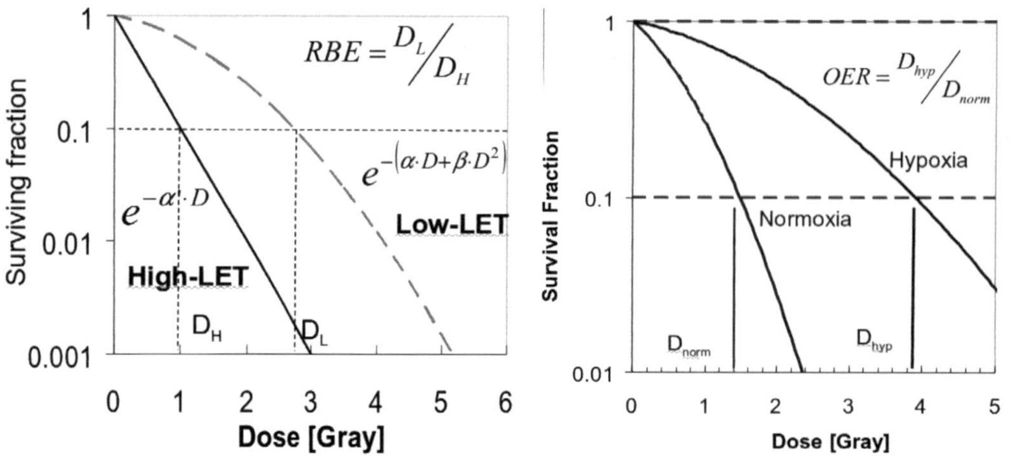

Figure 1. Illustration of cell survival curves. (Left) Effect of LET and definition of RBE; (Right) Effect of oxygenation and definition of OER.

survival at therapeutic doses shows linear dependency and a shoulder (figure 1). Practically, these curves for cell survival may be described by a linear quadratic model according to the following equation:

$$S(D) = exp(-\alpha D + \beta D^2) \tag{1}$$

where α and β are parameters, which are fitted using experimental data.

Compared to high-energy photons, high-LET ions have a twofold impact: firstly, cell killing may be more efficient, leading to larger values for α; secondly, β may vanish, leading to the disappearance of the shoulder on the curves. The *Relative Biological Efficiency* (RBE) is defined to help qualify the efficiency of a particular ionizing radiation in comparison to a reference one, typically high-energy photons: it corresponds to the ratio of the dose required with the reference radiation over the dose required with the studied irradiating particle, to induce a given endpoint. This coefficient is a quite complex function of many parameters, some biological (like the biological endpoint itself, the cell line and its environment), some physical. The latter are for instance the dose and the dose-rate, which, in the context of therapy, refer to the duration of each treatment session and the number of sessions (fraction) that are necessary to deliver the whole dose prescription. The dose-rate effects imply the dynamical response of cells to irradiations (e.g. DNA repair), and tissue recovery. The dependency of the RBE with respect to the type of ionizing particle cannot be reduced to the LET. A particle may induce less cell killing than another with lower LET (see for instance the results of our experiments at the GANIL/Irrabat facility [7]). Hence, the type of particle must be specified. For a fixed type of ions, the dependency of RBE with LET is typically the following: at low LET values, the RBE is close to unity, then it increases to a maximum and then decreases down to zero. This latter decrease is referred to as the "over-killing effect". Considering all types of ions, the maximal value of RBE is the highest with protons. However, this value corresponds to low-energy protons (of the order of 1 MeV), i.e. ions for which the range is too low to kill more than one cell. In therapy, the number of ions is quite low compared to the number of cells to kill. Therefore, the RBE must be high for ions fast enough to reach many cells in the tumor volume, but low in the

Journal of Physics: Conference Series **629** (2015) 012004 doi:10.1088/1742-6596/629/1/012004

healthy tissues. Carbon ions appear to be an interesting compromise. As an illustration, let us consider a volume of 120 cm^3 made of spherical cells each with radius 10 μm. The number of cells is therefore 6^{10}. An irradiation of this volume using carbon ions with 1 Gray requires an average of 7^8 incident carbon ions [8]. If this irradiation induces death in 50% of the cells present in this volume, this means that each carbon ion kills an average number of 50 cells. Even in the extreme scenario where one ion impact alone is sufficient to kill a cell, within this configuration, such "efficient" ions should range 1 *mm*, corresponding to the 18 MeV/n carbon-ion range in water. This rough estimation shows that the local control of tumors cannot be attributed to the highest-LET carbon ions, but relies on a compromise between a high value of RBE and the capacity for ions to cross many cells. It also explains why the average RBE in tumors irradiated with protons is close to 1, despite the large RBE value for MeV protons.

The Oxygen Enhancement Ratio (OER) refers to the oxygen effect. When irradiating cells at various levels of oxygenation, it has been observed that radiosensitivity increases with oxygenation up to a saturation for a partial pressure of oxygen lower than normoxy (normal oxygenation of tissue). In the same way as for RBE, the OER coefficient is defined for a given level of biological endpoint as the ratio of the dose required in hypoxia over the dose required with the tested level of oxygenation to induce this particular level of endpoint. The level of oxygenation in the cells of some tumors may be much lower than the one observed in normal conditions. The tumors in hypoxia become radioresistant thereafter. Such tumors are potentially indications for a boost with carbon ions, since the oxygen effect has been observed to decrease when the LET increases. In particular, close to the Bragg peak, the OER is as low as one.

3. Origin of RBE larger than 1

The origin of the biological efficiency of ions is still an open question. Somewhat arbitrarily, the basic scenarios have been divided into three classes.

3.1. Geometrical interpretation and microdosimetry

Historically, the first class refers to microdosimetry models and may be explained using the structure of ion-tracks. In the original framework of these models, the lethal lesions are induced by the interaction of sub-lethal lesions, with a typical distance of interaction in the order of one micrometer. As represented in figure 2a), the particularity of ion tracks leads to a spatial correlation in the distribution of sub-lethal events. In particular, when the ion LET increases, the distance between sub-lethal lesions decreases and therefore the probability for the occurence of lethal lesions increases. In practice, the predictions are based on microdosimetry calculations. Presently, the most refined version of microdosimetry models, the Microdosimetry Kinetic Model (MKM) [9], is used for clinical treatment in Japan, and several research teams are extending the validity of MKM to carbon therapy [10, 11]. The targets that are considered in microdosimetry are not clearly defined. The creation of sub-lethal lesions by damaging DNA is nonetheless mentioned. Considering the typical distance of efficient sub-lethal interactions, those implied in such models could involve fragments of DNA or even chromosomes. Two arguments at least support microdosimetry models. First, it has been shown that high-LET ions could induce complex chromosomal rearrangements [12]. Second, calculations based on Monte Carlo simulations [13] have shown that high-LET ions generate smaller DNA fragments than low-LET ones do.

3.2. Local dose deposition and track-structure models

In this class of models, the high capacity of ions to deposit very high local dose is at the origin of their biological efficiency. Using the concept of radial dose to estimate the local dose is a common feature in models of this class. Historically, this idea was supported by Katz *et al.* in proposing the γ-kill and ion-kill model [14]. The Local Effect Model (LEM) was then proposed by Scholz *et*

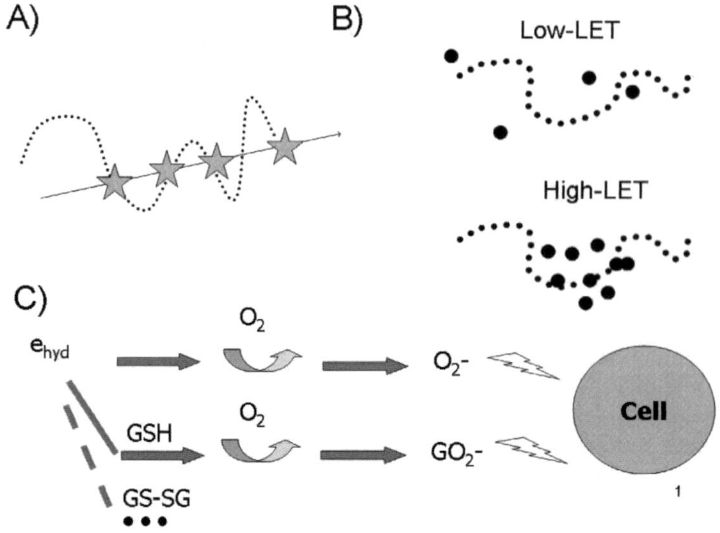

Figure 2. Illustration of the mechanisms based on a) the spatial correlation of sub-lethal lesions (stars) in DNA (dotted line); b) concentrations of radiation events (full circles) around DNA; c) direct or indirect interactions of hydrated electrons (or hydrogen radicals) with dioxygen molecules producing chemical species that are toxic for cells (illustrations of equation 3a and 3b).

al. [15]. In Katz's model, the target size is clearly defined to be of the order of the micrometer. In the LEM, the lethal events are considered as point-like, referring nevertheless to complex DNA damages (see figure 2b)). Both models are capable of predicting the main tendency of RBE with an acceptable accuracy. Both can also predict hook structures in the curves of inactivation cross section plotted as a function of ion LET. However, as pointed by Beuve *et al.* [16], the use of the radial dose concept in radiobiology has strong limitations regarding the stochastic nature of radiations, in particular for small targets. This raises questions in particular on the basic concepts of the LEM, as the target size is extremely small [17]. This conceptual problem led us to propose a new framework referred to as the Nanox model.

3.3. Core ionization

Alternatively, Chetioui *et al.* [18] pointed out a correlation between the cross-section for cell inactivation and the cross-section for core ionization when plotted as a function of LET (see figure 3). In particular, both show a section of increase with LET, followed by the hook structure. This idea was supported by an estimation of the efficiency for these core ionizations to induce lethal events, from a comparison of the cell killing induced by soft X-rays with photon energy below and above the binding energy of core electrons [19]. This efficiency, defined as the ratio of lethal-event number to the core ionization number, was estimated to 3-4 % [20]. Injecting this estimation into a Monte Carlo simulation led the authors to conclude that core ionization would represent 75 % of the lethal events produced by high-energy photons.

At this stage, this framework known as the model K, was not designed to allow cell survival prediction, but instead to allow discussions about the dominant mechanisms. Still, we propose

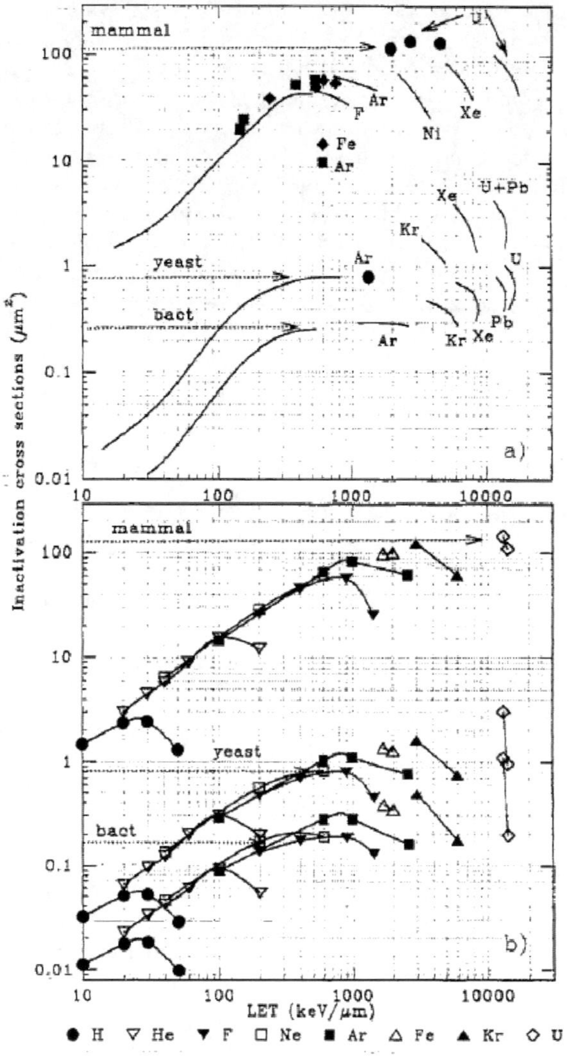

Figure 3. a) Experimental cross section for cell inactivation by heavy ions of various LET: mammal cells, yeast, bacillus subtilis; arrows indicate the geometrical cross section of the cell nucleus or of the DNA concentration. b) Calculated cross sections for the induction of at least one efficient K-vacancy in cell nucleus (from [21]).

to discuss about the application of the model K for hadrontherapy. Doing this raises at least two difficulties. The first one lies in the dose dependency of the lethal-event. Core ionization events are definitely localized events (atomic scale). These events, designated as lethal events in the model K, are therefore local lethal events. For such a class of events, it is possible to show that cell survival is a purely decreasing exponential (see for instance [17]). On the other hand, the measurements of cell survival show at least a linear quadratic dependency (as shown by the presence of a shoulder). This indicates that, in the framework of the model K, the efficiency ϵ to induce lethal events cannot be constant but should be a dose-dependent function $\epsilon(D)$. A second difficulty appears when comparing the efficiency of fast protons and carbon ions to induce lethal events and core ionizations. At a high but "fixed" ion velocity, both the LET and the core-ionization cross-section scale with the square of ion charge. Therefore, at a given fluence, the number of core ionizations on the one hand and the dose, on the other hand, also scale with the ion charge square. Thus, the number of core ionizations per unit of dose is the same for protons and carbon ions. In the context of the model K, one expects the number of lethal events per unit of dose to be similar for protons and carbon ions (with equal velocities). However, the experimental value of this ratio, which corresponds to the α parameter in the linear quadratic model of cell survival, may differ by a factor as high as 5 or more when comparing these two ion species. This indicates either that core ionization is not the main source of lethal events alone, or that the efficiency defined in the model K is likely to be a complex function of the dose, the LET and the type of ions. Knowledge of this function would be necessary to apply the model K to hadrontherapy. When comparing the model K prediction to the effective cross-section based on the slope of cell survival at high doses (and not at low doses), the discrepancy is reduced to a factor 2. This remaining factor is interpreted by the authors of model K in two ways. Firstly, by the multi-ionisation implying several core levels, which increase with the charge of incident ions and would induce more severe damage in DNA. Secondly, by events, referred to *K events*, that combine the high density of ionizations generated by the ions and the core ionizations, including Auger electrons.

4. Origins of the OER reduction

The reduction of OER to 1 at high LET is all the more under question since the origin of the radiosensitive effect of oxygen is still unknown. One scenario is based on the capacity of oxygen molecules to fix damage before their restoration by an antioxidant. The following reaction scheme (T standing for "Target") illustrates this scenario with hydroxyl radical attack ($OH^{\bullet}$) and glutathione antioxidant (GSH):

$$\begin{aligned} &\text{Target attack:} && TH + OH^{\bullet} \longrightarrow T^{\bullet} && (2a)\\ &\text{Restoration:} && T^{\bullet} + GSH \longrightarrow TH + GS^{\bullet} && (2b)\\ &\text{Fixation:} && T^{\bullet} + O_2 \longrightarrow TOO^{\bullet} && (2c)\\ & && TOO^{\bullet} + GSH \longrightarrow TOOH + GS^{\bullet} && (2d) \end{aligned}$$

In hypoxia, the fixation of damage would be inefficient and the damage capacity of radiation less important. As initially suggested by Neary [22], ions could produce oxygen molecules in their tracks, thus compensating the lack of oxygen in hypoxia. This production of oxygen has been observed and simulated [23, 24, 25, 26, 27, 28]. It is initiated by the multi-ionization of water molecules [28] and the reaction of the oxygen atoms issued with the chemical radicals of the track. However, as illustrated on figure 4, the production of oxygen in the track seems very low compared to the concentration of oxygen at which the oxygen effect saturates. This saturation appears at a partial oxygen pressure ranging from 3 up to 30 mmHg, while the normal partial oxygen pressure (normoxy) is 160 mmHg.

Alternatively, the production of superoxide anion ($O_2^{\bullet -}$ and its proton form $HO_2^{\bullet}$) is suspected to be implied in the origin of the oxygen effect. This superoxide anion is produced by the following reactions:

$$\begin{aligned} &\text{Direct interaction:} && e_{hyd} + O_2 \longrightarrow O_2^{\bullet -} && (3a)\\ &\text{Indirect interaction:} && e_{hyd} + GSH \longrightarrow G^{\bullet -} + HS^{\bullet -} && (3b)\\ & && G^{\bullet -} + O_2 \longrightarrow GO_2^{\bullet -} && (3c) \end{aligned}$$

This scenario is supported by chemical arguments based on the Fenton reaction and by biological evidence [30, 31, 32, 33, 34], in particular in relation with superoxide dismutase enzymes. Moreover, simulations of radical production by irradiation of water with various

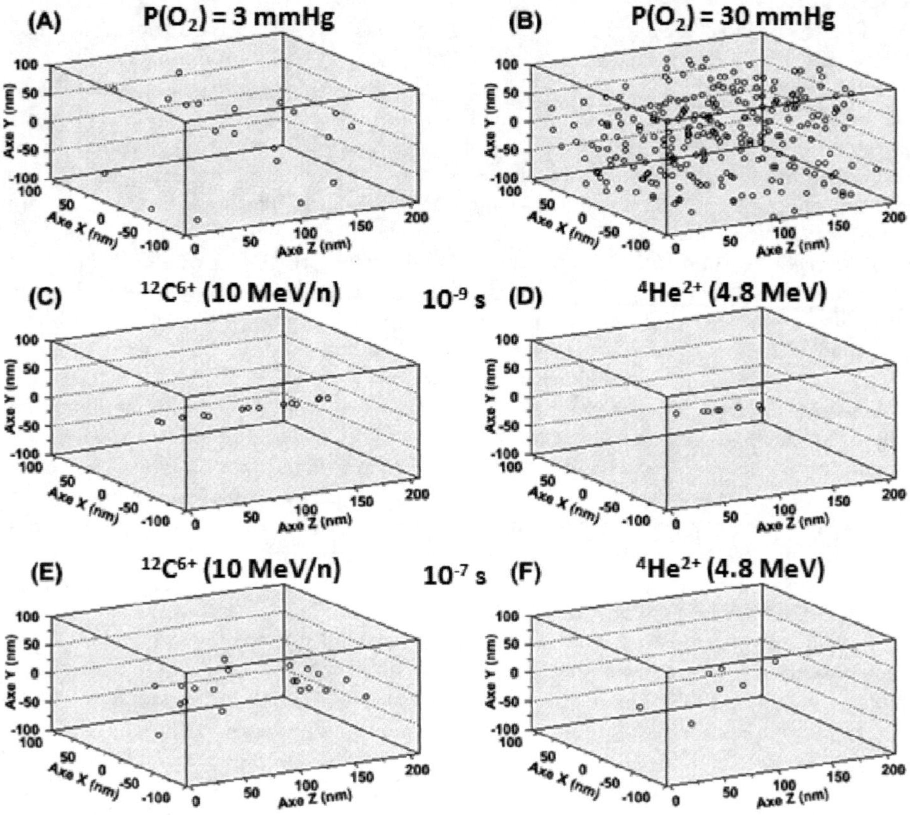

Figure 4. Spatial distributions of oxygen molecules dissolved in water at a partial pressure of 3 mmHg (A) and 30 mmHg (B) or produced in the track of C[10 MeV/n] (LET=168 keV/µm) and He[1.2 MeV/n] (LET=94 keV/µm) at 10^{-9} s C and D) and 10^{-7} s (E and F) after ion impact [29]. 3 mmHg and 30 mmHg correspond to the range of partial pressures at which OER begins to saturate for various cell lines.

oxygen concentration have shown a correlation between the oxygen effect and the production of $HO_2^{\bullet} + O_2^{\bullet -}$ when varying LET, particle type and oxygen concentration [35]. However, considering the high probability for hydrated electrons (e_{hyd}) and hydrogen radicals (which are at the origin of the production of $HO_2^{\bullet} + O_2^{\bullet -}$) to be scavenged before reacting with oxygen, Colliaux *et al.* suggested a more general scenario: although hydrated electrons (e_{hyd}) and hydrogen radicals may have a low capacity of damage, by reacting either directly equation 3a) or through intermediate molecules, as for instance glutathione equation 3b), these two radicals may lead to the production of secondary chemical species that could be toxic for cells either by damaging some structures or by impacting some cell signaling (see figure 2c)).

Journal of Physics: Conference Series **629** (2015) 012004 doi:10.1088/1742-6596/629/1/012004

5. DNA target

DNA is often considered as a crucial target in the mechanisms driving both the biological efficiency of light ions and the reduction of the oxygen effect. Although many other targets (and non-targeted events) should be considered, the DNA molecule is at least to be taken into account for such biological endpoints as mutation induction, chromosomal aberrations or instability and, to some extent, cell death. The detailed study of damage formation in DNA in cell appears quite complex both in *in vivo* and *in vitro* conditions. Alternatively, quantifying the damage generated by interactions of fast ions with biological elementary molecules, like sugar or DNA bases, in gas phase, appears quite simpler and more direct. However, such measurements may not provide enough information in the field of radiobiology. For instance, the effect of confinement induced by the molecule environment (mainly water in cells) may strongly modify the relaxation channels of the excited and ionized molecules: ionized molecules may fragment in the gas phase while they may find other modes of relaxation in liquid phase. Moreover, in a water environment, most of the damage in molecules is generated by the free radicals produced by water radiolysis. This mechanism, referred to as indirect action, cannot be ignored. Another limitation to the knowledge that these experimental systems may provide to the radiobiology domain is related to the geometry of these elementary molecules. Spotheim-Maurizot *et al.* [36, 37] have shown that the geometrical structures of DNA molecules influence the signature of damage, due to the accessibility of free radicals to the sites of DNA. The sequence of nucleotides, but also the different levels of DNA compaction, should be considered when characterizing damage in DNA. Nevertheless,the quantification of the mechanisms induced by the interaction of ions with molecules in the gas phase may provide data to challenge complex models of the early interaction stage, like the quantum *ab initio* description of collision and molecule relaxations.

As an alternative target to elementary molecules in the gas phase, DNA plasmids in water constitute a step toward DNA in cells. With such systems, E. Sage *et al.* have demonstrated the capacity of high-LET ions to create complex damages. Known as clustered DNA lesions or locally multiply damaged sites (LMDS), these are defined as two or more damages in the DNA molecule within 1-2 helical turns [38]. Among them, the dirty double-strand break (DSB) [39], defined as a combination of a double-strand break close to a single base lesion, appears difficult to be handled by cells. All these complex damages increase the risk for mutation induction and cell death triggering.

Studies with plasmids have also shown that other elements played a role in the mechanisms of damage formation. The protection by antioxidants, that are able to scavenge the radiation-induced radicals before their interaction with DNA, reduces the number of damage [42]. The enzymatic activity modifies the yield and the signature of DNA damage, by repairing some lesions but also by inducing others. Urushiba *et al.* [43] have studied DNA damage induction by irradiating plasmids in solutions containing two types of enzymes (the endonuclease III (Nth), a base excision repair enzyme, and the formamidpyrimidine DNA glycosylase) at various concentrations. They showed that enzymatic activity could modify the yield of damage. They also confirmed the quite robust tendency that increasing LET decreases the yield of single-strand breaks (SSB) but increases the yield of DSB. These examples indicate that studies of DNA damage formation in cells are necessary to include all the mechanisms triggered in actual biological systems, a far from simple task in practice (see for instance [44]).

A question addressed in many papers concerns the relative contribution of direct and indirect actions to DNA damage formation. Often quoted, the pioneering work by Roots *et al.* [45] states that indirect actions dominate at low LET, while direct actions dominate at high LET. This work uses strong concentrations of hydroxyl radicals in scavenger (DMSO) and assumes that these concentrations are as high in the nucleus. Besides the fact that this methodology is questionable, we stress that Ito *et al.* have shown with the same protocol that indirect actions dominate for LET as high as $440 keV/\mu m$ [46]. In studying the signature of base damage, J-L Ravanat *et al.* [47] have shown that the indirect damage of DNA bases is also dominant at high LET. By modulating the concentration of glutathione in tumor cells, a major antioxidant in cell nucleus, Hanot *et al.* [48] have shown that a depletion of glutathione induce more cell killing and more residual DSB. Considering the high capacity of this antioxidant to scavenge free radicals, these observations are in favor of an important contribution of indirect actions for high-LET ions as well. Finally, the question on the relative contribution of direct and indirect actions still is opened and requires more experimental evidence.

6. Conclusion

Interaction of ions with biological systems is an active domain of research with an important impact on society, like the research for a better estimation of ionizing-radiation risks for health, or the improvement of cancer treatments by carbon ion therapy, an innovative therapy that benefits from the particular physical and biological properties of fast carbon ions. This latter opportunity was born from the results obtained through many years of research activities, combining modeling and experiments (performed with high-energy accelerator facilities, like GANIL) and many disciplines and actors, including physicists, chemists, biologists and physicians. Despite the numerous results, many fundamental or more clinically-oriented questions require clarification. As an illustration of this scientific and societal requirement, the national Infrastructure "France Hadron" aims at coordinating researches related to hadrontherapy and facilitating access to ion beams. This initiative will contribute to the increase of knowledge in such a complex but interesting field, which involves the interaction of high-energy ions with matter.

Acknowledgments

Several results presented in the present paper have been supported by the Rhône-Alpes Region Programme for Research in Hadrontherapy (PRRH) and the GDR MI2B, and/or, have been performed within the framework of the LABEX PRIMES (ANR-11-LABX-0063).

References

[1] Wilson R R *et al* 1946 *Radiology* **47** 487
[2] Min C H, Kim C H, Youn M Y and Kim J W 2006 *Appl. Phys. Lett.* **89** 183517
[3] Testa E, Bajard M, Chevallier M, Dauvergne D, Le Foulher F, Freud N, Létang J M, Poizat J C, Ray C and Testa M 2008 *Appl. Phys. Lett.* **93** 093506
[4] Henriquet P, Testa E, Chevallier M, Dauvergne D, Dedes G, Freud N, Krimmer J, Létang J M, Ray C, Richard M H and Sauli F 2012 *Phys. Med. Biol.* **57** 4655
[5] Gwosch K, Hartmann B, Jakubek J, Granja C, Soukup P, Jäkel O and Martišíková M 2013 *Phys. Med. Biol.* **58** 3755
[6] Parodi K, Bortfeld T and Haberer T 2008 *Int. J. Radiat. Oncol., Biol., Phys.* **71** 945
[7] Beuve M, Alphonse G, Maalouf M, Colliaux A, Battiston-Montagne P, Jalade P, Balanzat E, A, Bajard M and Rodriguez-Lafrasse C 2008 *Int. J. Radiat. Oncol., Biol., Phys.* **71** 635
[8] Krämer M, Jäkel O, Haberer T, Kraft G, Schardt D and Weber U 2000 *Phys. Med. Biol.* **45** 3299
[9] Hawkins R B 2003 *Radiat. Res.* **160** 61
[10] Inaniwa T, Furukawa T, Kase Y, Matsufuji N, Toshito T, Matsumoto Y, Furusawa Y and Noda K 2010 *Phys. Med. Biol.* **55** 6721
[11] Inaniwa T, Suzuki M, Furukawa T, Kase Y, Kanematsu N, Shirai T and Hawkins R B 2013 *Radiat. Res.* **180** 44
[12] Testard I and Dutrillaux B S L 1985 *Radiat. Res. Suppl.* **8** S112–2
[13] Friedland W, Dingfelder M, Kundrát P and Jacob P 2011 *Mutat. Res. Fund. Mol. Mech. Mut.* **711** 28
[14] Katz R, Ackerson B, Homayoonfar M and Sharma S C 1971 *Radiat. Res.* **47** 402
[15] Scholz M, Kellerer A, Kraft-Weyrather W and Kraft G 1997 *Radiat. Environ. Biophys.* **36** 59
[16] Beuve M, Colliaux A, Dabli D, Dauvergne D, Gervais B, Montarou G and Testa E 2009 *Nucl. Instrum. Meth. Phys.* B **267** 983
[17] Beuve M 2009 *Radiat. Res.* **172** 394
[18] Chetioui A, Despiney I, Guiraud L, Adoui L, Sabatier L and Dutrillaux B 1994 *Int.J. Radiat. Biol.* **65** 511

Journal of Physics: Conference Series **629** (2015) 012004 doi:10.1088/1742-6596/629/1/012004

[19] Touati A, Hervé du Penhoat M A, Bailly-Despiney I, Gobert F, Champion C, Fayard B, Abel F, L'Hoir A, Moulin J, Sabatier L and Chetioui A 2002 *Radiat. Prot. Dosim.* **99** 83
[20] Boissière A, Champion C, Touati A, Hervé du Penhoat M A, Sabatier L, Chatterjee A and Chetioui A 2007 *Radiat. Res.* **167** 493
[21] Touati A, Hervédu Penhoat M, Bailly-Despiney I, Gobert F, Champion C, Fayard B, Abel F, L'Hoir A, Moulin J, Sabatier L and Chetioui A 1997 *Nucl. Instrum. Meth.* B **132** 276
[22] Neary J G 1965 *Int. J. Radiat. Biol.* **9** 477
[23] Sauer M C, Schmidt K H, Jonah C D, Naleway C A and Hart E J 1978 *Radiat. Res.* **75** 519
[24] Burns W G, May R and Baverstock K F 1981 *Radiat. Res.* **86** 1
[25] Baverstock K F and Burns W G 1981 *Radiat. Res.* **86** 20
[26] Baldacchino G, Bouffard S, Balanzat E, Gardés-Albert M, Abedinzadeh Z, Jore D, Deycard S and Hickel B 1998 *Nucl. Instrum. Meth.* B **146** 528
[27] Gervais B, Beuve M, Olivera G H, Galassi M E and Rivarola R D 2005 *Chem. Phys. Let.* **410** 330
[28] Gervais B, Beuve M, Olivera G H and Galassi M E 2006 *Radiat. Phys. Chem.* **75** 493–513
[29] Colliaux A 2009 *Implication de l'oxygéne et des antioxydants dans le processus de radiolyse de l'eau induite par l'irradiation aux ions de haute énergie : simulations numériques pour la radiobiologie*, PhD thesis, Université de Lyon (France)
[30] Aikens J and Dix T A 1991 *J. Biol. Chem.* **23** 15091
[31] Goldstein A and Czapski G 1995 *Free Radic. Biol. Med.* **19** 505
[32] DeGrey A 2002 *DNA Cell. Biol.* **21** 251
[33] Mikkelsen R B and Wardman P 2003 *Oncogene* **22** 5734
[34] Delattre J, Beaudeaux J L and Bonnefont-Rousselot D 2005 *Radicaux libres et stress oxydant. Aspects biologiques et pathologiques* (Éditions Lavoisier, Paris, Prance)
[35] Colliaux A, Gervais B, Rodriguez-Lafrasse C and Beuve M 2011 *J. Phys. Conf. Series* **261** 012007
[36] Spotheim-Maurizot M, Ruiz S, Sabattier R and Charlier M 1995 *Int. J. Radiat. Biol.* **68** 571
[37] Begusova M, Spotheim-Maurizot M, Sy D, Michalik V and Charlier M 2001 *J. Biomol. Struct. Dyn.* **19** 141
[38] Sage E and Harrison L 2011 *Mutat. Res. Fund. Mol. Mech. Mut.* **711** 123
[39] O'Neill P and Wardman P 2009 *Int. J. Radiat. Biol.* **85** 9
[40] Milligan J R, Aguilera J A, Nguyen T T, Paglinawan R A and Ward J F 2000 *Int. J. Radiat. Biol.* **76** 1475
[41] Fulford J, Nikjoo H, Goodhead D T and O'Neill P 2001 *Int. J. Radiat. Biol.* **77** 1053
[42] Hada M and Sutherland B M 2006 *Radiat. Res.* **165** 223
[43] Urushibara A, Shikazono N, O'Neill P, Fujii K, Wada S and Yokoya A 2008 *Int. J. Radiat. Biol.* **84** 23
[44] Gault N, Rigaud O, Poncy J L and Lefaix J L 2007 *Radiat. Res.* **167** 551
[45] Roots R, Chatterjee A, Chang P, Lommel L and Blakely E A 1985 *Int. J. Radiat. Biol.* **47** 157
[46] Ito A, Nakano H, Kusano Y, Hirayama R, Furusawa Y, Murayama C, Mori T, Katsumura Y and Shinohara K 2006 *Radiat. Res.* **165** 703
[47] Douki T, Ravanat J L, Pouget J P, Testard I and Cadet J 2006 *Int. J. Radiat. Biol.* **82** 119
[48] Hanot M, Boivin A, Malésys C, Beuve M, Colliaux A, Foray N, Douki T, Ardail D and Rodriguez-Lafrasse C 2012 *PloS one* **7** e44367

From ion-hammering to ion-shaping: an historical overview

Giancarlo Rizza

Ecole Polytechnique, Laboratoire des Solides Irradiés, CEA-IRAMIS-CNRS, 91128 Palaiseau Cedex, France

E-mail: giancarlo.rizza@polytechnique.edu

Abstract. The objective of this article is to summarize the research activity carried out within the field of amorphous materials submitted to swift heavy-ion irradiation since the opening of the GANIL facilities. In particular, this review focuses on two experimental observations which have created excitement in the swift heavy-ion community for several years: the ion-hammering and the ion-shaping processes. However, this article is neither a comprehensive review nor a tutorial. Indeed, its prime purpose is to give an historical overview of these two phenomena, putting the accent on how a new phenomenon is addressed by the scientific community and how through a difficult trial-and-error approach unsolved problems favor the issue of a coherent description of the phenomenon.

1. Introduction

I was asked by the committee for the *30 years of interdisciplinary research at GANIL* to write an historical review of the study of the effects induced in glassy systems by swift heavy ions. In particular, amongst the large amount of existing experimental results I was asked to overview two phenomena: the anisotropic plastic deformation and the elongation of embedded nano-particles along the the beam direction. However, instead of giving a account of the experimental facts and theoretical predictions in chronological order, I will take this opportunity to describe how, in my humble opinion, a new phenomenon is addressed by the scientific community.

The explosive development of a new research field is almost always associated with the development of a new instrumental technique that allows new ideas to be checked. It is easier to study a phenomenon if the proper instrumentation to observe it is available. However, posing the right questions and giving the right answers is not always the path researchers follow when they are facing a new phenomenon. Usually, a complex and slow trial-and-error process will after years of research activity converge toward a coherent, but often not unanimously accepted, description of this new observation within the scientific community. In my opinion there exists an heuristic recursive schema that can be generally applied to describe how a new observation is first treated, then integrated and finally exploited by the researchers. This process can be schematically summarized into three main steps:

i) the dawn of a new physics or of the required technical and social background conditions,

ii) the discovery of a new phenomenon and the first steps toward a phenomenological description of it,

iii) from phenomenological to quantitative models and potential applications.

In the following, the ion-hammering and the ion-shaping phenomena will be reviewed within this empirical framework.

Journal of Physics: Conference Series **629** (2015) 012005 doi:10.1088/1742-6596/629/1/012005

2. The historical background

The physics of swift heavy ions is tightly related to the physics of ion tracks, which in turn is the physics of how the energy deposited by an impinging charged particle is transferred to both electronic and atomic sub-systems.

The first experimental observations of the ion tracks in materials dates back to the late fifteens with the works of Young [1] and Silk and Barnes [2]. These seminal works opened the door to a stimulating research activities in a variety of fields spanning from nuclear physics, to radiation dosimetry and interplanetary science. Earlier observations pointed out that an ion track is mainly due to inelastic processes, i.e. the radiation-induced excitation and ionization of the target atoms. However, owing to the limited power of the existing accelerators, these studies were mostly restricted to insulators and organic materials. Meanwhile, two different models were proposed to account for the existing experimental results. A model based on the concept of a Coulomb explosion [3], and a model based on the concept of a thermal spike [4]. The former assumes that the track core is always solid and that shear stress, whose magnitude approaches the ideal shear strength, is generated. The latter assumes that the track core is transiently fluid. The issue which one of the two approaches is more adapted to describe the track formation has a longstanding history that has its importance in this historical overview. At this point of this historical overview, what is possible to say is that till the early 1980s, the Coulomb explosion model was considered as the more appropriate one. The reason is threefold: i) at that time the thermal-spike models were not sufficiently developed to provide useful predictions, ii) the experimental fact that it is difficult to generate tracks in materials having a large Young's modulus was nicely accounted for by the criterion that tracks form only when the stress resulting from Coulomb explosion surmounts the yield strength of the material and iii) it was implicitly assumed that the mechanism leading to the track formation does not hold in electric conductors.

As mentioned above, science often progresses by technological steps and the physics associated to swift heavy ion irradiation does not make an exception to this rule. Although, since its discovery in the early 1960s, an important deal of effort has been devoted to rationalizing the physics underlying the track formation, a real breakthrough took place around 1980, with the advent of large accelerator facilities (GANIL-Caen, GSI-Darmastadt and VICKSI-Berlin). Indeed, the access to these new large-power accelerators allowed the researchers to plan systematic experiments in regions of much higher excitation and ionization densities. This technological progress led to the dawn of the novel research activity described hereafter.

3. The discovery of the ion-hammering effect

The reason why tracks were not formed in metals was still an open issue in the early 1980s. It was often claimed that it was due to the high density of free electrons that quickly spread out the deposited energy and screen in a very short time the charge of the ionized atoms along the ion path. It is thus natural that, as soon as the large accelerator facilities became available to the research community, these materials were among the first ones to be investigated. Among them a particular focus was put onto metallic glasses. Indeed, they were seen as suitable radiation-resistant materials, as their disordered structure was supposed to prevent the irradiation growth by easily absorbing the radiation-induced disorder.

It was therefore with surprise that the first irradiation experiments on unstressed metallic glasses revealed large macroscopic deformations, figure 1a). The first experimental observation was reported in 1983 by Klaumünzer *et al* in a $Pd_{80}Si_{20}$ metal glass irradiated with 360 MeV Kr ions at the VICKSI accelerator in Berlin [5]. The irradiated sample was observed to undergo macroscopic changes at constant volume and without appreciable modifications in its microscopic structure. It was quite soon evident that this unknown phenomenon was different from all the deformation processes already observed in crystals, i.e. swelling, irradiation creep and irradiation growth. However, in analogy to the dimensional changes observed in anisotropic metals, see e.g. [6], it was first named *growth*. At this point, it is worth noticing that often the name given to a new phenomenon changes as the understanding of the phenomenon evolves. As the growth mechanisms that are active in crystalline and amorphous metals are totally

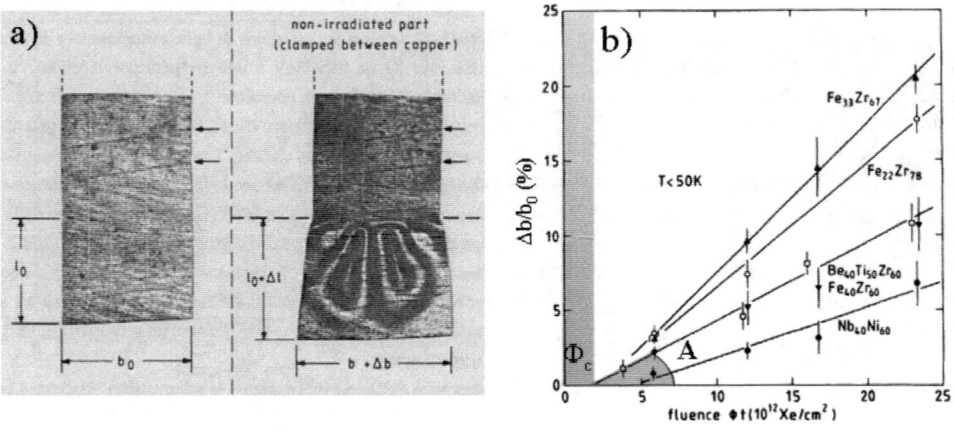

Figure 1. a) micrographs of (left) an non-irradiated $Co_{75}Si_{15}B_{10}$ metallic glass and (right) the same sample where the lower part has been irradiated at 50 K with 360 MeV Xe ions up to $1.7\text{x}10^{13}$ cm^{-2}. b) relative change in sample dimension for different metallic glasses. Adapted after reference [9].

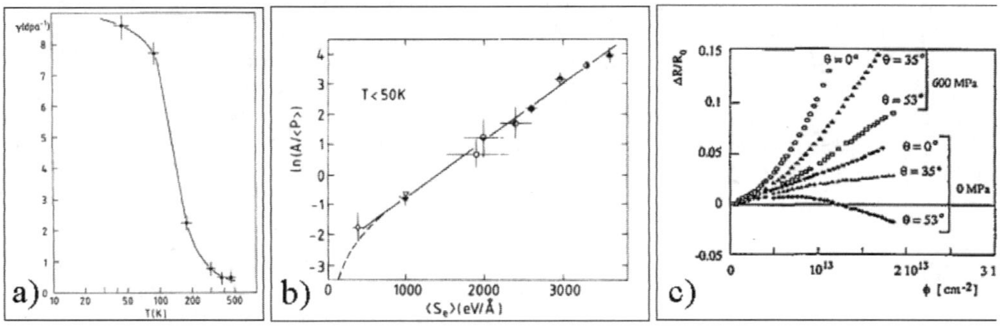

Figure 2. Dependence of the ion-hammering effect on the irradiation conditions: a) irradiation temperature, T, b) electronic stopping power, S_e and c) external stress, σ. Adapted after a) [5], b) [7], c) [10].

different, the name was subsequently changed to *anisotropic deformation* or *plastic flow* and finally to *ion-hammering*. The latter terminology conveniently highlighted the role played by each impinging ion: i.e. sample deforms in a manner as each ion would act like a little hammer.

3.1. The first experimental results: searching for a key parameter

During the ion-hammering process the sample dimensions perpendicular to the ion beam, *b*, are observed to grow whereas the sample dimension parallel to the ion beam, *a*, are observed to shrink in such a way that the mass density remains unaltered. In particular, it was observed that, above an incubation fluence, Φ_C, the strain, $\varepsilon=\Delta b/b$, linearly scales with the irradiation fluence, Φ, such that the following empirical relationship holds:

$$\frac{\Delta b}{b_0} = A\,(\Phi - \Phi_C) \tag{1}$$

where b_0 is the initial width of the sample. The coefficient A represents the strain rate, or the rate at which the ion-hammering occurs. It is characterized by the slope of the time-dependent deformation curve, figure 1b), and it is defined as $\dot{\varepsilon}=d\varepsilon/d\Phi$. In the literature it has also been called the *growth rate* or the *deformation yield*.

In an early paper, Klaumünzer pointed out that equation (1) can be also obtained by considering the continuity equation $\dot{\rho}+\rho\vec{\nabla}\cdot\vec{v}=0$, where ρ is the mass density and v is the velocity of any sample point [7].

Quite soon, the deformation yield became the key parameter to unveil the mechanisms underlying the ion-hammering phenomenon. Indeed, it provided a powerful tool to qualitatively categorize the large amount of available experimental results [5, 7–12], figure 2a-c). Indeed, it allowed to show that the growth rate was directly related to the electronic stopping power, S_e. Moreover, it was shown that the deformation yield depends on irradiation temperature, on the electronic stopping power and the stress state of the target. The deformation yield is maximal at low temperature and decreases as soon as thermally activated atomic rearrangement are possible. Below the track generation threshold, A is small and decreases roughly exponentially with decreasing S_e. Above the track generation threshold, A increases linearly with with S_e. Moreover, the deformation yield increases linearly with an externally applied stress. Finally ion-hammering was not observed in materials which remain crystalline during irradiation.

However, the study of the ion-hammering did not remain limited to metallic glasses. Indeed, soon after its discovery, it became clear that the ion-hammering was an universal phenomenon occurring not only in metallic glasses but in all amorphous materials, [12], such as dielectrics like SiO_2 [13] or semiconductors like a-Si [14]. Again, the behavior of the amorphous material upon swift heavy ion irradiation could be classified on the basis of the deformation yield. It has been observed that A roughly scales with the glass transition temperature and the high-temperature coefficient of linear thermal expansion but not with the elastic properties of the materials [15].

3.2. The first steps toward a phenomenological description

Early experiments pointed out that the deformation yield was directly related to the electronic stopping power. Thus, the first phenomenological interpretations turned out to be focused on how local order modification is efficiently transformed into a macroscopic deformation.

At that time the dominating model for track formation was the Coulomb explosion model. Thus, the first phenomenological interpretations were (*cognitively*) biased by this approach [7, 9, 10]. For example, Hou *et al.* stated that "*the constitutive atomic rearrangements are probably shear transformations induced by transient stresses which attend the ion explosion spikes in the wake of the fast ions*" [9]. Audouard *et al.* proposed a two-hit model where the driving force for the atomic motion is the pressure induced by a shock wave. Thus, the macroscopic deformation results from the accumulation of the damage produced by the ion-induced compression waves [10].

In 1990 Hou *et al.* suggested that microscopically each ion acts like an hammer [9]. The macroscopic deformation is due to the accumulation of local plastic strains triggered by energy deposited by the ion into the electronic sub-system. The authors proposed to define the deformation yield per ion as: $A=\varepsilon\pi R^2$, where ε is the irreversible local radial strain and πR^2 the track cross-section. This simple relationship helps to qualitatively quantify the ion-hammering induced by a single ion impact. As the deformation yield, A, is in the range 10^{-14}-10^{-15} cm^2, for a track radius of R=3 nm, it follows that the $\varepsilon=10^{-2}$-10^{-13}. Besides, considering that the deformation time is of the order of 1ps, the strain rate must be in the range 10^9-10^{10} s^1.

In 1991 Klaumünzer and Benyagoub published the first phenomenological model describing the ion-hammering phenomenon [16], figure 3. The authors used continuum mechanics to describe the dimensional changes of the sample upon irradiation by a strain tensor. In particular, they succeeded in

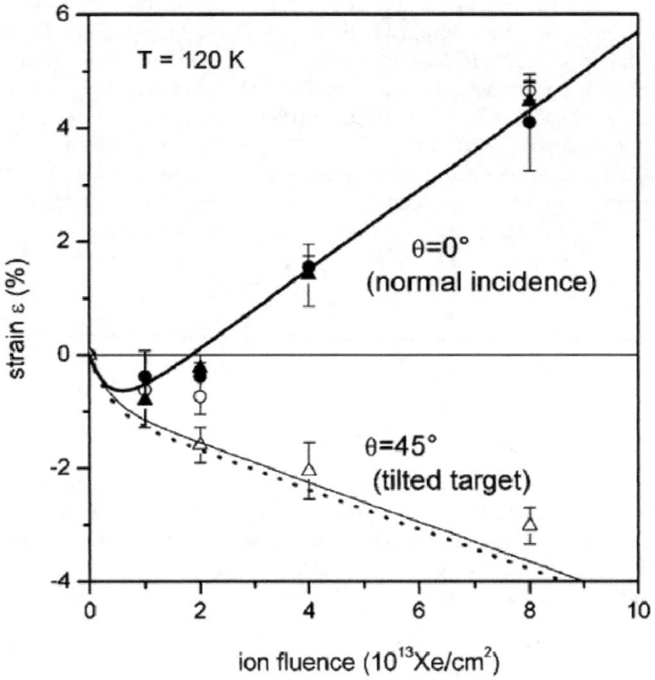

Figure 3. Dimensional changes of a vitreous silica thin film irradiated at 120 K with 360 MeV Xe ion. The experimental data for the ion-hammering are recorded for two irradiation geometries: i) θ=0° and ii) θ=45°. Adapted after [13].

accounting for the ion-hammering when the irradiation angle was changed. In a cartesian coordinate system in which the z-axis is along the sample normal and the y-axis coincides with the tilt axis the deformation tensor, $\overline{\overline{A}}$, reads:

$$\overline{\overline{A}} = A(T, S_e, \sigma) \begin{pmatrix} 1-3sin^2\theta & 0 & 3sin\theta cos\theta \\ 0 & 1 & 0 \\ 3sin\theta cos\theta & 0 & 1-3cos^2\theta \end{pmatrix} \quad (2)$$

where A(T,S_e,σ) denotes the deformation yield per ion.

3.3. From phenomenological to quantitative models

The first attempts to interpret the ion-hammering process were based on the Coulomb explosion model, e.g [7, 17]. However, its predictive power remained quite limited and none of these phenomenological approaches was able to give an analytical expression for the deformation yield. Besides, it was observed that when mechanical stress is present, creep rates scale linearly with the applied stress [10, 18, 19]. Thus, the radiation creep in amorphous materials must be described as a Newtonian flow with an effective fluidity proportional to the ion flux. As the shock mechanism is beyond linear elasticity, Coulomb explosion approach turned out to be incompatible with a creep phenomenon which is linear in stress [20]. Meanwhile, using Auger and convoy electron spectroscopies, Schiwietz *et al.* demonstrated that in

Journal of Physics: Conference Series **629** (2015) 012005 doi:10.1088/1742-6596/629/1/012005

metals electrical space charges are virtually neutralized within femtoseconds such that the occurrence of a repulsive Coulomb force of sufficient strength and lifetime is an unlikely process [21–26].

With the time the interest of the SHI community for this approach therefore gradually faded way and the thermal-spike models underwent a renaissance. This change of paradigm was especially due to Toulemonde and Dufour [27] and Szenes [28]. In the thermal-spike models the target solid is divided into two subsystems: the electronic (e) and the atomic (a) ones. The model states that the energy of the projectile is deposited into the electronic subsystem where it becomes thermalized within 10^{-15} sec. The energy is then transferred to cold atoms through an electron-phonon interaction, and a thermal equilibrium is reached after about 10^{-13} sec. Within this framework, the thermal evolution of both the electronic and the atomic subsystems can be described by numerically solving the following coupled differential equations:

$$C_e(T_e)\frac{\partial T_e}{\partial t} = \frac{1}{r}\frac{\partial}{\partial r}\left[rK_e(T_e)\frac{\partial T_e}{\partial r}\right] - g\,(T_e - T_a) + A_e(r,t) \tag{3}$$

$$C_a(T_a)\frac{\partial T_a}{\partial t} = \frac{1}{r}\frac{\partial}{\partial r}\left[rK_a(T_a)\frac{\partial T_a}{\partial r}\right] + g\,(T_a - T_a) \tag{4}$$

Within the thermal-spike approach, the ion-track is defined as the cylindrical region exceeding the melting temperature of the material. This useful definition allowed theoretical predictions and experimental observations to be compared directly. This is probably the main reason why the Coulomb explosion approach was discarded in favor to the thermal-spike approach.

However, thermal-spike models alone were not sufficient to give an insight into the ion-hammering phenomenon. Indeed, the critical issue was tightly related to the development of a relationship relating the ion-track to the build-up of a localized strain state around it. Otherwise stated, an analytical description of the deformation yield, A, was necessary. The main problem toward this goal was the definition of an amorphous track in an amorphous material. Indeed, if in a crystal the concept of track is intuitively clear, how to describe an amorphous track in an amorphous material? This problem was solved in a series of papers published in the period 1995-1998 first by Trinkaus and Ryazanov [29], then by Trinkaus alone [30–32].

The underlying idea was to use Eshelby's classical theory of elastic inclusion [33], to describe an ion-track as a visco-elastic inclusion. Nowadays, this approach is known as the effective-flow-temperature approach (EFTA) and the basic ideas of the model are described hereafter:

The formation of an ion-track is associated to the melting of a cylindrical region around the ion path, whose extension is characterized by the flow temperature T^*. Stresses are generated by the thermal expansion of this cylindrical molten region. The EFTA model states that the principal stresses in the direction normal and parallel to the ion beam are different, i.e. $\sigma_{11}=\sigma_{22}\neq\sigma_{33}$. Thus, a shear stress emerges. The latter can be completely relaxed if its life-time, τ_s, is shorter than the duration of the thermal spike, τ_{TS}. However, the duration of the thermal-spike process is shorter than the shear stress relaxation time, such that upon rapid cooling the strain increment freezes-in. The resulting strained track represents the basic mesoscopic defect responsible for the ion-hammering. The overall effect is due to the sum of all the frozen-in viscous strains produced by the individual ion impacts.

The EFTA model gives an analytical expression for both the deformation yield, A, and the steady-state the in-plane stress, $\sigma_\perp$:

$$A = 0.427\frac{1+\nu}{5-4\nu}\frac{\alpha S_e'}{\rho C_L} \tag{5}$$

$$\sigma_\perp = -1.16\frac{1+\nu}{1-\nu}G_\infty\alpha(T^* - T_0) \tag{6}$$

where α is the coefficient of linear thermal expansion, S_e' the fraction of the stopping power that has been converted into heat, ρ the target density, ν the Poisson number and G_∞ the high-frequency shear

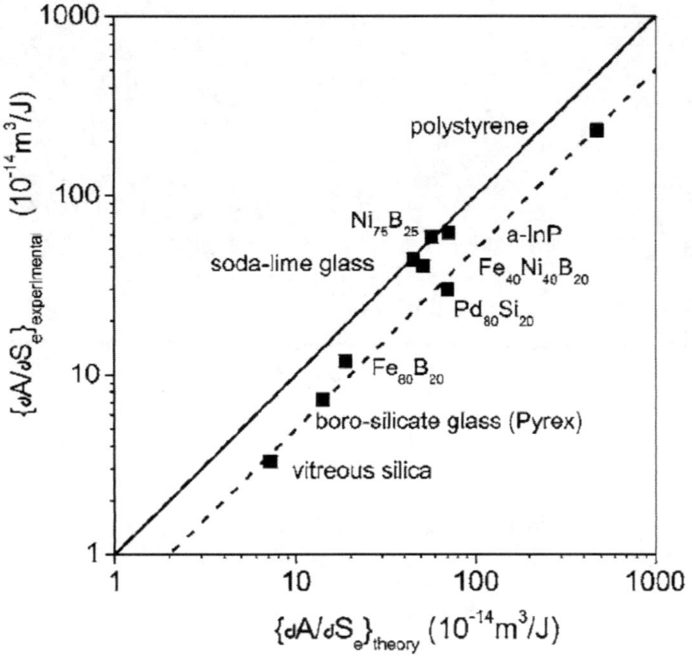

Figure 4. Comparison between the experimentally determined slopes of the deformation yield $\partial A/\partial S_e$ and the predicted theoretical values calculated from equation 5 using S_e'=f(S_e-S_{e0}), with f=0.5 (dashed line) and f=1 (solid line). Adapted after [11].

modulus of the liquid. However, in the EFTA model instantaneous shear stress relaxation occurs for $T>T^*$ and instantaneous freezing for $T<T^*$. The incorporation the time-dependence of the relaxation process is due to van Dillen *et al.* in a paper published in 2006 [34].

The analytical expression for the deformation yield allowed experimental values and theoretical estimations to be directly compared [11, 35, 36]. One example is given in figure 4 where the experimental and the theoretical slopes of the deformation yield, i.e. $\partial A/\partial S_e$, are clearly correlated for a wide variety of amorphous materials.

3.3.1. Constitutive equations for the ion-hammering. In the first version of the EFTA model, the shear stress is instantaneously relaxed, conversely frozen, at T^*. To avoid this unphysical assumption, in 1998 Trinkaus succeeded in obtaining the complete constitutional equation describing ion hammering at low temperature [11, 32]:

$$\frac{d\varepsilon_{ij}}{dt} = A_{ij}\frac{d\Phi}{dt} + \frac{1}{2\mu}\frac{d}{dt}\left[\sigma_{ij} - \frac{\nu}{1+\nu}\sigma\sigma_{kk}\delta_{ij}\right] + \frac{s_{ij}}{2\eta} \tag{7}$$

The first term describes the unconstrained anisotropic deformation process. The second term represents the time-dependent elastic behaviour of the material, i.e. it corresponds to Hooke's law for isotropic elastic media, where σ_{ij} are the non-hydrostatic capillarity stresses and μ and ν the shear

Journal of Physics: Conference Series **629** (2015) 012005 doi:10.1088/1742-6596/629/1/012005

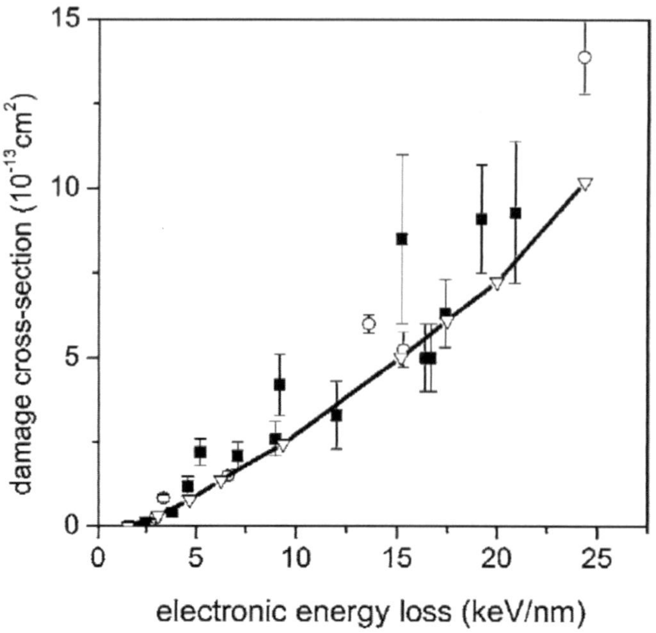

Figure 5. Damage cross-sections as a function of S_e for quartz and vitreous silica. Full squares and open circles are experimental results, whereas open triangles and solid line represent the model calculations according to Eq. 7. Adapted after [11].

modulus and the Poisson ratio, respectively. The third term describes the inalastic behavior of the material, i.e. the macroscopic irradiation-induced Newtonian viscous flow. Here, s_{ij}=(σ_{ij}-1/3 $\sigma_{kk}\delta_{ij}$ are the deviatoric stresses and $\eta\propto(d\Phi/dt)^{-1}$ is the ion-flux-dependent shear viscosity.

For example, the use of equation (7) allows to reproduce the damage cross-section in quartz and vitreous silica as a function of the stopping power S_e, figure 5.

4. The irradiation science strays into the nano-world

A change of the viewpoint in the swift heavy ion community started to develop since the late nineteens with the growing interest of the scientific community for the nanoworld: *there's plenty of rooms at the bottom* [37].

The emergence of nanotechnology started in the early 1980s mainly driven by two factors. On one hand, the publication of the Drexler's book "*Engines of creation: the coming era of nanotechnology*", who helped to popularize the concept of nanoscience as a mean to assemble the matter at the nanoscale [38]. On the other hand, two breakthrough technical advancements: the invention of the scanning tunneling microscope in 1981 and that of the atomic force microscope in the same year. Under the impulse of this scientific excitement, other techniques and approaches to the nanoscale went into a rapid growth process. Just to mention a few examples: the continuous development of the techniques associated to the transmission electron microscopy, photo-lithography, electron-beam lithography, and colloidal chemistry.

The swift heavy ion community was hit by this new scientific wave starting from the late 1990s. Indeed, if a lot of work has been done to understand the ion-matter interaction in the regime of pure

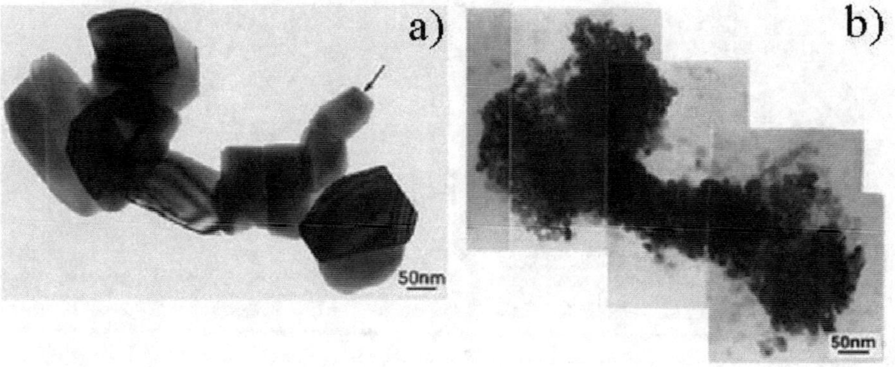

Figure 6. Tin oxide grains: (a) non-irradiated sample, and (b) after irradiation with 4.6 MeV.u^{-1} Pb ions at a fluence of 5×10^{12} cm^{-2}. The arrow indicates a small grain which partially disappears after irradiation. Adapted after [40].

electronic energy deposition, most of the experiments have been performed either on macroscopic samples or polycrystalline samples with a mean grain size exceeding a micrometer. At that time, it became clear that one way to obtain a deeper insight into the swift-heavy ion-matter interaction was to reduce the size of the target. Among these pioneering works, we can mention the cubic to monoclinic phase transformation observed in Y_2O_3 sub-micrometric powders [39] and the irradiation of Tin oxide nanograins [40], figure 6a-b).

4.1. The ion-hammering at the nanoscale

In a series of seminal papers published between 2000 and 2006, Polman's group demonstrated that the ion-hamming was not only limited to macroscopic samples but it was effective even at the nanoscale, figure 7. In particular, exploiting the development of the colloidal chemistry, Snoeks *et al* [41, 42] and van Dillen *et al.* [43–46], extended the work of Cartz *et al.* to sub-micrometer silica particles [47]. Indeed, in 1981 Cartz et al. discovered that ion-amorphized silicate micro-particles (<5 μm) undergo an irradiation creep process when irradiated with 3 MeV Kr ions.

The value of these works is twofold. On one hand, this novel approach allowed to investigate the ion-hammering at the nanoscale. On the other hand, it introduced a novel paradigm in the study of the swift-ion irradiation phenomena. The ion-beam became a tool to sculpt the matter at the nanoscale paving the way to more applied research topics (the fabrication of track-etched membranes is another example of this novel approach). The combination of chemical synthesis and ion-irradiation techniques rendered it possible to produce a new class of colloidal ellipsoids of continuously variable shape. Soon after the publication of these results, it was demonstrated that the ion-hammering of colloidal silica nanoparticles was successfully described by the EFTA model [49].

We have seen that for an unconstrained amorphous thin film, the experimental data can be modeled by introducing the rate-of-strain tensor, e.g. $A^0_{ij} = d\varepsilon_{ij}/d\Phi$. The solution of this equation in the direction perpendicular to the ion beam is an exponential function of the fluence:

$$L(\Phi) = L_0 Exp(A_0\Phi). \qquad (8)$$

This equation states that the ion-hammering is an insaturable process, figure 8b). However, when the thin film is replaced by a nanoparticle, capillarity stresses, which are related to the curvature radius of the nanoparticle, must be considered. This corresponds to the second and third terms of equation (7).

Journal of Physics: Conference Series **629** (2015) 012005 doi:10.1088/1742-6596/629/1/012005

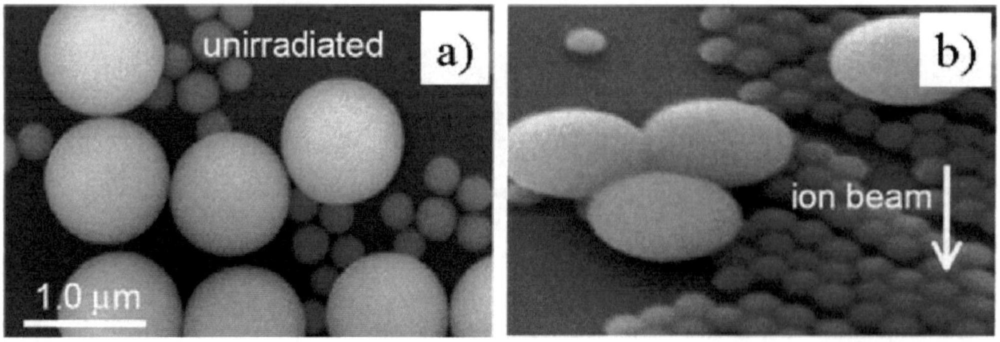

Figure 7. Scanning electron microscopy images of a) non-irradiated silica colloids and b) after 4 MeV Xe irradiation at a fluence of 3×10^{14} cm^{-2}, at an angle of 45° relative to the surface normal, at 90 K. Adapted after [41].

For large nanoparticles, capillarity forces are small and the deformation kinetics is still given by Eq. (8). However, as pointed out by Klaumünzer [49] when the curvature radius is reduced, capillarity stresses becomes important and the two previous terms must be considered. As the signs for the ion-hammering and for the capillarity forces are opposite, ion hammering should be, in principle, balanced by viscous flow. The slowing down of the ion-hammering process when the colloidal size was reduced was first put in evidence by van Dillen *et al* [46], figure 8c). In 2011 Rizza *et al.* [50] showed that the deformation yield at the nanoscale is not longer constant but that it becomes a function of the irradiation fluence, i.e. $A \longrightarrow A(\Phi)$. Furthermore, they show that it vanished for large irradiation fluences, i.e. $A \longrightarrow 0$. Otherwise stated, the ion-hammering saturates, figure 8a-c).

5. The discovery of the ion-shaping

Motivated by these novel observations, D'Orléans, Stoquert and collaborators decided to study how the properties of embedded metallic nanoparticles were modified when submitted to a sustained swift heavy ion irradiation. In line with the approach developed by Polman's group, the ion beam was used as a tool to modify the properties of embedded nanoparticles. Indeed, in a series of articles published over 2003-2004, D'Orleans *et al.* [51–53] reported that Co nanoparticles embedded in a silica matrix changed shape, from spheres to rods, under the influence of swift heavy-ion irradiation. Interestingly, while van Dillen's experiments show that the silica colloids become oblate, i.e. the deformation is normal to the beam direction, in D'Orléans's experiments the metallic nanoparticles become prolate, i.e. the major axis of the elongated nanoparticles was aligned parallel to the incident ion direction. The ion-shaping process was discovered.

This surprising finding has rapidly attracted the attention of an important number of groups worldwide [54–87]. The effect has been reproduced in many systems, and in general the results indicate that within a wide range of metals, the nano-particles can become elongated under suitable irradiation conditions.

It is interesting to note that in the first paper of D'Orléans *et al* [51] the deformation of the Co nanoparticles was not particularly highlighted in the title: "*Irradiations of implanted cobalt nanoparticles in silica layers*". However, in their following paper the term of *anisotropic deformation* was introduced [52]: *Anisotropy of Co nanoparticles induced by swift heavy ions*. The same years, the same terminology was adopted by Roorda *et al* [80] and Penninkhof *et al* [82]. In 2007, Schmidt et al [77] first introduced the term *ion-shaping* to describe the ion-deformation of crystalline Ge NPs embedded in a silica matrix. Nowadays, this term has been accepted by the community and currently indicates the irradiation-induced transformation of spherical nanoparticles into prolate nanostructures (e.g. [61, 62]).

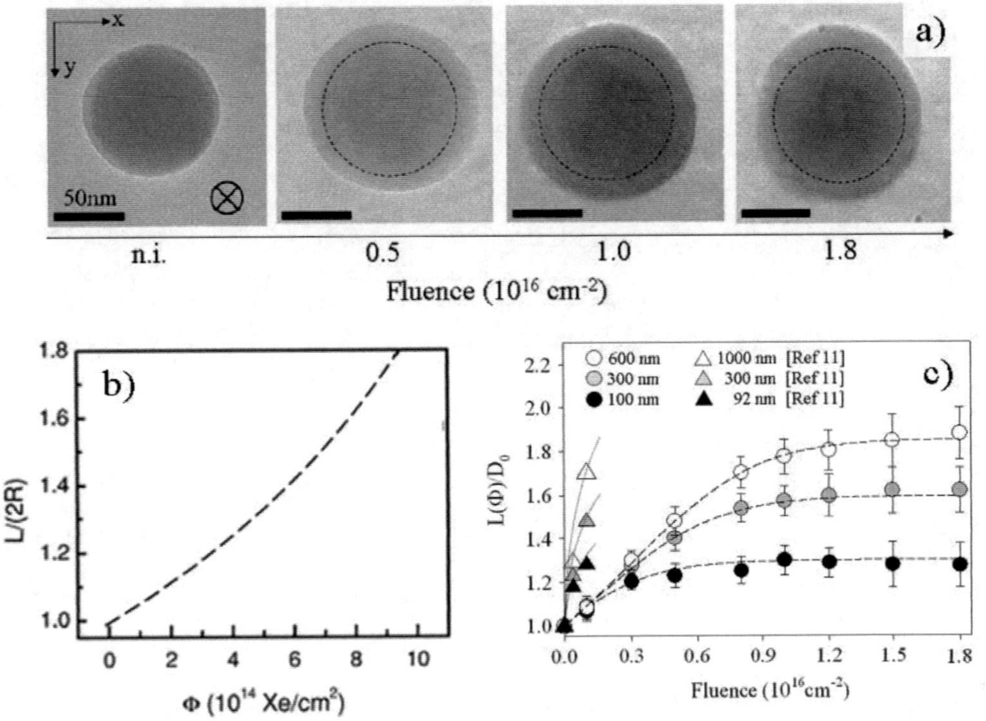

Figure 8. a) Bright-field micrographs showing the deformation of colloidal silica NPs (100 nm) as a function of the irradiation fluence. The ion beam is normal to the plane, ⊗. b) The dashed line shows the exponential growth behavior for free anisotropic plastic deformation given by equation 8 with $A_0=6.2\times10^{-16}$ cm^{2}ion^{-1}. c) Evolution of the normalized transverse diameter of the NPs $(L\Phi)/D_0$ as a function of the ion fluence. D_0 is the NP diameter. Circles are experimental data from [50] and triangles from [46]. Adapted after a) [50], b) [46] and c) [50].

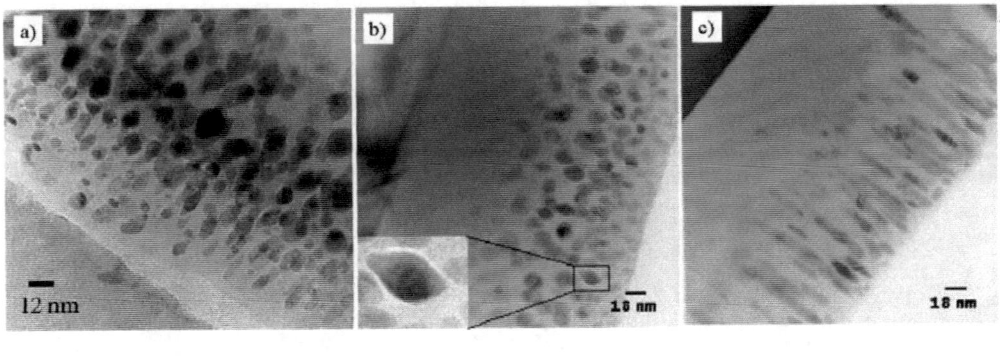

Figure 9. Ion-shaping of Co NPs embedded in silica matrix. Adapted after a) [51], b) [52] and c) [52].

Hereafter, the experimental and theoretical developments will be described separately. Although this method of presenting the subject is somewhat arbitrary, it allows to focus on the difficult journey followed by the researchers to reach a coherent description of the phenomenon. We could describe this journey as a succession of ephemeral truths. First a brief summary of the experimental knowledge acquired over the last decade on the ion-shaping phenomenon will be given. Then, the intellectual efforts made to obtain an insight into the ion-shaping phenomenon will be described. In fact, in the present stage of knowledge, a coherent theoretical description of the ion-shaping process is still lacking.

5.1. The accumulation of experimental results

The fact that the deformation yield is a key parameter in the ion-hammering process emerged rapidly. However, it less obvious that there exists a similar unique key-parameter for the ion-shaping process. To put it simply, the ion-shaping is basically the transformation of a sphere of diameter D into a cylinder of width W and length L. Thus, a great deal of effort has been made to correlate these quantities to both the irradiation conditions and the thermodynamic properties of both the matrix and the embedded NPs. In the first part, the studies on nanocomposite thin films will be summarized. In the second part we will briefly describe he research activity on core-shell colloidal NPs.

5.1.1. Ion-shaping in nanocomposite thin films Right from the early experiments on, it has become clear that the ion-shaping depended on the NP dimension [51, 52, 63, 64]. In particular, three different deformation regimes were identified. NPs smaller than about 8-10 nm do not deform. However, they are observed to grow under irradiation, most likely through a process akin to Ostwald ripening where larger particles grow at the expense of smaller ones [51, 52, 64, 65]. NPs in the range 10-20 nm deform into nanorods and nanowires whose main axis is aligned with the direction of the beam. Finally, until 2009 the consensus in literature was that NPs larger than about 20 nm could not be deformed upon irradiation. It is however worth noticing that their deformation had been clearly observed in TEM micrographs, but not recognized as such, see e.g. figure 3a) in [51], figure 4 in [66] or figure 2 in [67]. The previous statement was justified by the fact that the thermal-spike simulations indicated that the threshold size for melting was about 20 nm [51, 63].

It was not until 2009 before the fact that ion-shaping was not limited to NPs smaller than 20 nm became explicitly noticed [62, 64, 68]. In particular, they were observed to evolve toward facetted-like morphologies [68].

At this point one is really entitled to wonder why a whole community had accepted for several years the paradigm that NPs larger than 20 nm were not deformable despite the contrary evidence in the published TEM images which clearly indicated the opposite. The source of the problem has a twofold origin.

The first reason is inherent to the technique used to prepare the pristine sample, i.e. the ion-beam synthesis (IBS) technique. While IBS is well suited to fabricate nanocomposite materials, the broad size distribution of the embedded NPs makes it difficult to give a clear correlation between the initial NP size and its final morphology. Within the framework of the ion-shaping phenomenon, this problem was solved by Rizza *et al.* by introducing a novel fabrication method where nearly mono-dispersed colloidal nanoparticles are sandwiched between two silica layers [69]. By using this novel approach, the correlation between the initial size of a NP and its final morphology could be clearly established, figure 10. This allows to show that four different size-dependent regimes exist during the ion-shaping process:

- NPs smaller than about 10 nm remain spherical in shape.
- NPs in the range 10-30 nm transform continuously into nanorods and nanowires.
- When the NP size is further augmented (30-70 nm), an increasing refractoriness to deformation is observed. These NPs take longer times to be ion shaped and reveal a tendency to evolve toward facetted configurations.
- Finally, larger NPs (>70-80 nm) are not noticeably deformed even at the highest irradiation fluence.

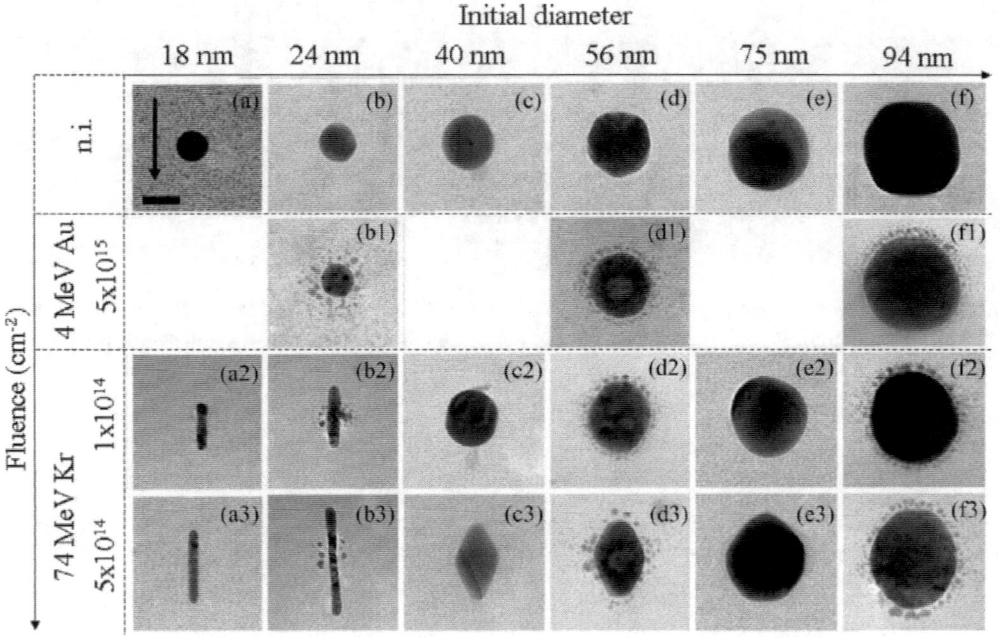

Figure 10. Matrix-like diagram showing the relationship between the initial NP size (x axis) and the irradiation fluence (y axis). The length of the scale bar is 20 nm. The ion-beam direction is indicated by the arrow. Adapted after [69].

The second reason is that dominating models may cognitively bias the interpretation of the experimental results. We have already seen this at work in section 3.2 for the application of Coulomb explosion model to the interpretation of the ion-hammering effect. The reason is that with time the dominant becomes more and more simply accepted on the basis of trust (this is especially true for the experimentalists). This point will be described in more detail in the next section.

Returning to physics, it was observed that for NPs larger than the threshold size, the ion-shaping proceeds through two regimes [62, 70]: i) for low irradiation fluences, the ion-shaping is an individual process where each spherical NP progressively transformed to a rodlike shape of fixed width independent of length. However, when the width of the nanorod/nanowire approaches a critical value, morphological instabilities appear and the NP becomes subject to breakup [70, 71] and dissolution [72]. The rate of dissolution increases with ion energy or equivalently with electronic stopping power, demonstrating that the dissolution process is not mediated by ballistic interactions. ii) for large irradiation fluences, the maximum length of the unbroken NWs has been observed to depend on the initial density of the embedded NPs, indicating that a collective growth process is active where more than one NP contributes to the elongation process. These experimental observations indicate that when the inter-particle distance is sufficiently small to permit solute diffusion upon irradiation between, the single NP deformation must be considered as a simplistic approximation and a more complicated dynamical process must be taken into account.

The fragmentation of the elongated NPs is related to the minimum sustainable width W of the NW and can be interpreted in terms of a Rayleigh-like instability under irradiation [70]. Ridgway *at al* have shown that the maximum width of an elongated NP scales with the electronic stopping power, S_e [61, 64, 74], i.e. it is limited by the diameter of the molten ion track within the host matrix. On the other

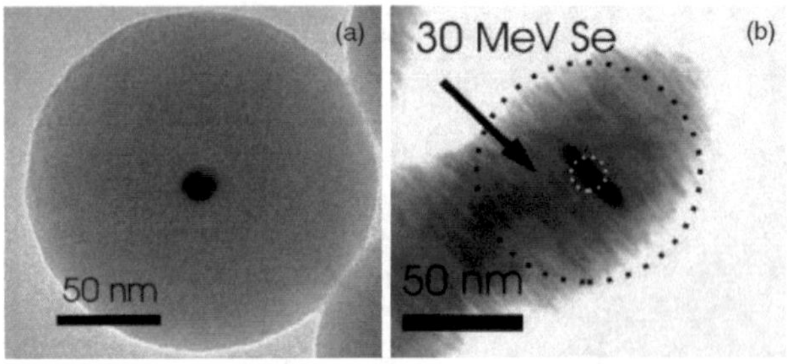

Figure 11. TEM micrographs of a) non-irradiated gold core (14nm) surrounded by a 72nm thick silica shell, b) after irradiation with 30MeV Se ions at an angle of 45° to the normal at a fluence of 2×10^{14} cm^{-2}. Adapted after [80].

hand, the minimum width is governed by the energy density per atom required for vaporization of the metal [74].

Systematic investigations on both the energy, S_e^c, and the fluence, Φ_c, thresholds have been performed by Dawi *et al.* using Rutherford backscattering (RBS) spectrometry [75]. It is deduced that above S_e^c, the deformation rate linearly scales with the electronic stopping power, S_e, and inversely scales with the NP size such that three size-dependent sub-domains can be identified: i) for Au NPs in the range 15-30 nm S_e^c is found to be between 3 and 4 keV.nm^{-1}. Thus above the value of S_e~2 keV.nm^{-1} necessary for the silica matrix to be deformed upon irradiation [13, 35]. ii) S_e^c is observed to increase up to 5.5-6 keV.nm^{-1} when the dimension of the NPs is increased to 40-50 nm iii) Finally, for 80 nm NPs the threshold is observed to range between 7 and 8 keV.nm^{-1}. The threshold fluence for deformation, Φ_c, has been observed to range between 10^{13} and 10^{14} cm^{-2} [52, 62, 63, 75], i.e. close to the fluence necessary to complete the compaction of the silica matrix [35]. However, Amekura et al. have shown by optical analysis that the ion-shaping can be induced without overlaps of ion tracks, thus below Φ_c [76]. Furthermore, the amplitude of the process has been observed to decrease with the deposited energy, S_e, and to increase with the NP size [62, 75].

Finally, Ge NPs embedded in SiO_2 show either prolate or oblate deformation depending on the NP size [77–79].

5.1.2. Ion-shaping of core-shell nanoparticles. In parallel to the investigation of the ion-shaping process in nanocomposite thin films, the study of the ion-shaping in metallo-dielectric core-shell colloidal micro-to-nanoparticles has developed, motivated by the explosive development of the field of of plasmonics.

In a series of articles published over 2004-2006 [80–82] it is shown that in Au-core/silica-shell colloids, the silica shell expands perpendicular to the ion beam, while the metal core deforms along the ion beam direction, provided the silica shell thickness is large enough ($>$ 40 nm), figure (11). Besides, metal deformation was observed to occur above an electronic energy loss threshold in the silica of about 3.3 keV.nm^{-1}.

The negative counterpart, i.e. silica-core/gold-shell colloids, was also studied [83, 84]. Here, the anisotropic deformation was attributed to the ion-hammering of the silica core, counteracted by the mechanical constraint of the metal shell. At constant fluence, an increase of the metal shell thickness reduces the deformation, while the addition of an extra silica shell results in deformation rates that approach the deformation rates of silica.

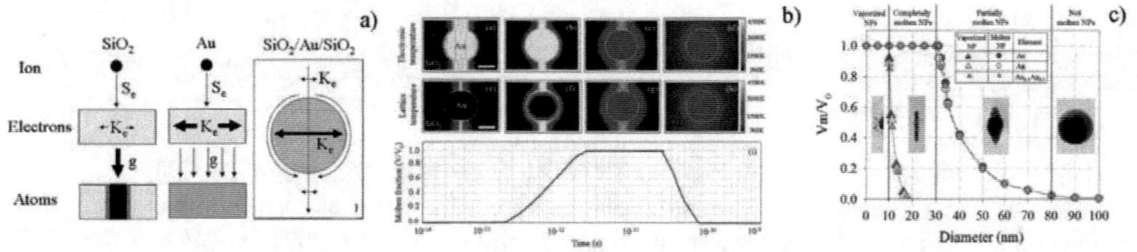

Figure 12. a) Schematic diagram for the swift-heavy-ion-matter interaction for (a) bulk silica, (b) bulk gold, and (c) gold NPs confined within a SiO_2 matrix. b) Time-line of the thermal evolution for a 20 nm Au NP embedded within a SiO_2 matrix irradiated with a 74-MeV Kr ion. Simulations have been run considering an ion hitting the NP at the pole. Time evolutions are shown for the electronic temperature, the lattice temperature and the fraction of the NP that reach a temperature exceeding the melting tempertaure, i.e. the molten fraction. c) The size-vs-shape diagram used to rationalize the ion-beam-shaping process for all the NP sizes. Four temperature-dependent regions can be defined (see text). Adapted after [69].

5.2. Toward a phenomenological description

Since its discovery, the ion-shaping has been extensively studied by many groups worldwide. Below we discuss the early discussions of the elongation mechanism, although no definitive mechanism to explain it has been proposed up to now.

The explanation of a new phenomenon is always a complicate task and the first attempts are often done in the light of the knowledge already acquired elsewhere. The case of the ion-shaping has not been an exception to this rule. The first phenomenological model was proposed by D'Orléans *at al* in 2003 [51, 52] and it is largely based on the idea proposed by Berthelot *et al* [40] in their study on the fragmentation of tin oxide (SnO_2) powder irradiated with swift heavy ions. That is, when the energy deposited by a swift heavy ion is sufficient to melt an embedded NP large shear forces are generated. The latter are due the thermal expansion of the molten NP that is constrained by the surrounding solid matrix. In particular, simulations reported in the D'Orléans paper [52], indicate the existence of a temperature gradient between the insulating matrix and the embedded NP. The latter generates a thermal pressure which drives the elongation of the molten NP through a creep mechanism. As described in the following, this idea was recently revisited and refined by Leino *et al* [85].

On the basis of the results obtained in core-shell systems, Roorda et al. proposed in 2004 an indirect deformation mechanism where the melting of the NPs is not considered [80]. Here, the ion-shaping is driven by the in-plane mechanical stress, that is generated by the ion-hammering of the silica shell, and acts on the radiation-softened Au NP. This mechanism favors the flow of the metallic species towards the out-of plane direction, i.e. along the direction of the ion beam. This mechanism has been extended to nanocomposite thin films.

The explanations proposed by D'Orléans *et al.* and by Roorda *et al.* are based on two completely different scenarios. In the first one, the melting of the metallic NPs is a necessary condition for the ion-shaping to be active. In the second one, the privileged mechanism is the creep of a solid (radiation-softened) NP enhanced by an in-plane mechanical stresses.

In a paper published in 2006 Klaumünzer tried to answer the question whether or nor the melting of the NP is an inevitable conclusion for the ion-shaping [86]. First, deformation mechanisms based dislocations or grain boundary diffusion were rejected. Then, the velocity of radial shrinkage of a void, representing the most deformable passive cluster, i.e. 1.4×10^{-13} ms^{-1}, was compared to the experimentally measured radial shrinkage for a cobalt NP, which is about 10^{-12} ms^{-1}. The larger shrinkage rate measured for the metallic NPs indicates that the passive deformation mechanism must be discarded and that the NP must play an active role in the deformation process, i.e. it must melt and it is thus subjected to a large thermal pressure.

Interestingly, in this paper the following statement is reported *"Particles with dimensions larger than 20 nm are expected not to melt"*. We have already mentioned that this assertion was in contrast with the available experimental results but in agreement with the simulation results based on the thermal-spike model. As pointed out by Klumünzer [11] although the existing TS models were successfully used to describe a large amount of SHI experiments, nonetheless they were based on a series of drastic approximations. Among them, three can be mentioned in this context: till 2008 i) the simulation of the thermal evolution of composite systems was not possible. In particular, due to the used boundary conditions, the diffusion of the heat through the metal/dielectric interface was not allowed. ii) TS models were based on a bi-dimensional approximation, and iii) in dielectric matrices, target electrons were considered to behave like a free electron gas. However, this approximation is only valid when the temperature exceeds the temperature which corresponds to the band-gap energy according to $T > T_g = k_B / E_g$.

To make progress towards a coherent description of the ion-shaping phenomenon it was necessary to overcome these limitations .

In 2008 Awazu *et al.* extended the thermal-spike model to account for the thermal evolution of a nanocomposite material [63]. The latter is simulated by two cylindrical regions centered at the ion trajectory, where the inner one represents the metal nanoparticle, and the outer one the silica matrix. The energy of the swift heavy ion is first deposited into the electronic sub-system of the metal nanoparticle. Then, it diffuses to peripheral electrons by electron-electron interactions and to the lattice by electron-phonon coupling, crossing the materials boundary at the metal-silica interface. The model maintains all the approximations used in the standard thermal-spike model [27]. Thus, only two-dimensional simulations are performed where only the heat flow in the plane normal to the ion beam is taken into account. However, spatial energy distribution has been modified to account for the presence of two materials around the ion path. The model tried to find a correlation between the ion-shaping and the melting of the nanoparticles. The main result is that the melting (and thus the shaping) only occurs for NPs smaller than 25 nm. Despite the improvement, the code failed thus in accounting for the deformation of large NPs.

The main problem with the *classical*, two dimensional, TS models is that they only provide a binary solution for the thermal evolution of the irradiated NP. Indeed, above 10 nm, NPs can only be either in a solid or in a molten phase. However, the reality has turned out to be slightly more complicated.

In this sense, a real breakthrough has been achieved by Khomenkov, Dufour and collaborators, [87]. Indeed, in a paper published in 2012, they introduced several major improvements to the existing TS codes. The first one concerns the implementation of the code in three dimensions and for a real nanocomposite material, i.e. the code allows to simulate all the NP shapes, where only coaxial cylinders were allowed in the Awazu's model. Secondly, following the work of Daraszewicz and Duffy [88], they extended the TS model to insulators. Indeed, the number of electrons involved in the thermal process, n_e, is considered to be proportional to the electronic temperature, T_e as long as T_e is lower than the band-gap temperature, T_g. Thirdly, the heat diffusing through the matrix/metal interface is explicitly taken into account.

Abandoning the oversimplified idea that the electrons in insulators behave like in metals and allowing the heat to diffuse through the matrix/metal interface, has proved to be the key parameters to unveil the mechanism behind the ion-shaping process observed for larger NPs.

A phenomenological description of the physics behind the simulation results is obtained by using the schematic diagram of figure 12a). Silica possesses a low electronic thermal conductivity, K_e, and a high value for the electron-phonon coupling constant, g. Thus, the energy deposited remains spatially localized within the ion trajectory where it is rapidly thermalized, resulting in a narrow and hot ion track. In contrast to silica, gold possesses a high value for the electronic thermal conductivity and a weak electron-phonon coupling. Thus, the deposited energy is swiftly smeared out over the electronic

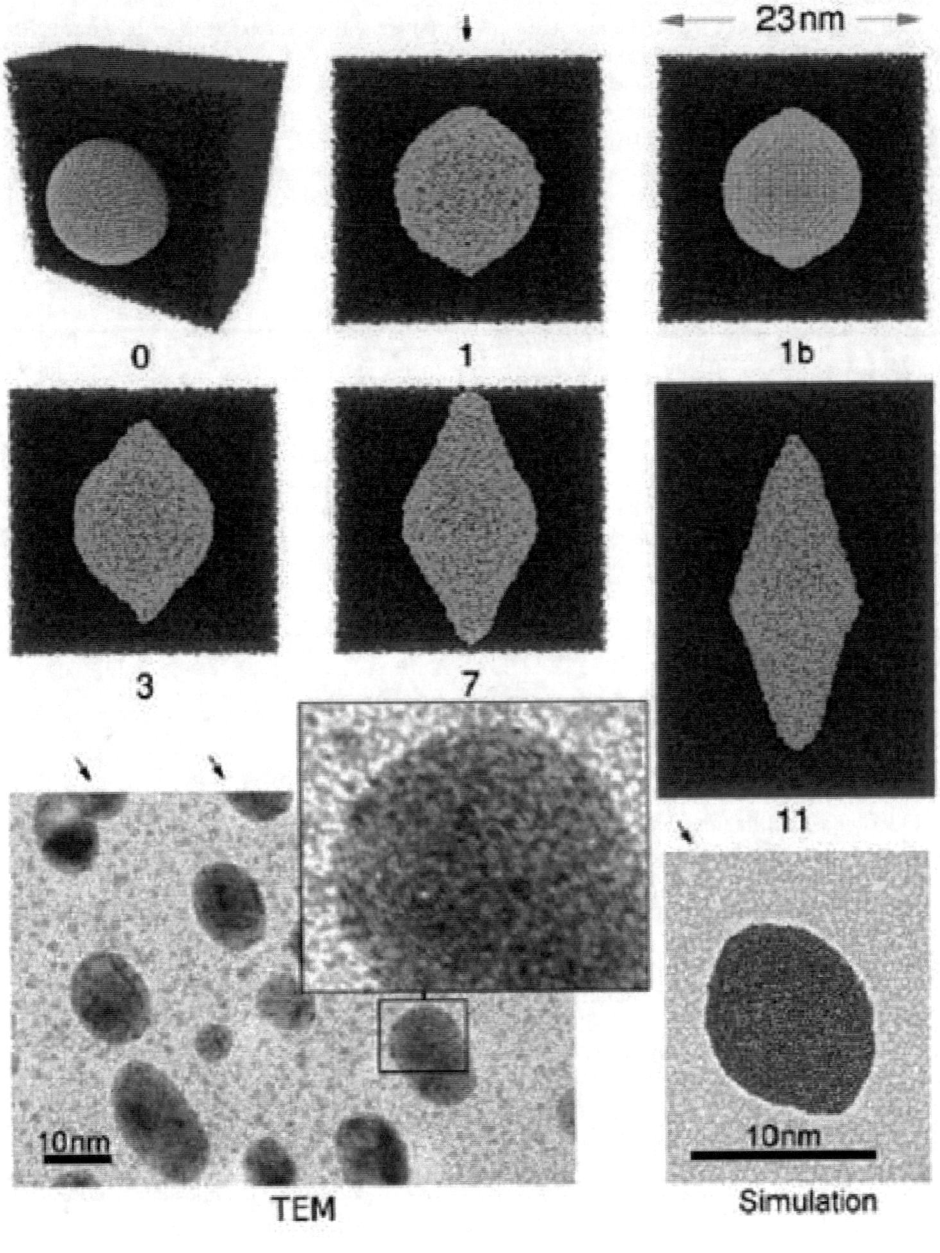

Figure 13. Elongation of Au nanocrystals in silica from Molecular Dynamic simulations. Each image shows a cross section of the simulation cell 50 ps after the ion impacts. The lower left image shows an experimental TEM image at an intermediate dose, and the lower right image shows the simulated shape after three non-overlapping impacts. Adapted from [85].

subsystem while it is only weakly coupled to the atomic subsystem. This results in a rapid increase of the electronic temperature but in a limited augmentation of the lattice temperature. The case for the Au NPs embedded within a silica matrix leads to an interesting effect. The energy transferred to the electrons of a Au NP diffuses rapidly outwards to the surface. When this energy reaches the Au/SiO_2 interface, it is transferred to the electrons of the neighboring silica matrix. Here, the large electron-phonon coupling results in the formation of a hot silica layer. Thus, the dielectric matrix acts as a barrier for the diffusion of the electronic energy, allowing, at the same time, its transformation into heat. Then, in a further stage the heat diffuses back toward the center of the NP. Thus, we can conclude that the irradiation-induced heating of a gold NP (high K_e, low g) confined within a silica matrix (low K_e, high g) can be seen as an indirect mechanism wherein both the NP and the surrounding matrix participate in the process. Due to this indirect-heating mechanism, the temperature profile within the NP becomes strongly size dependent, figure12b). In particular, complete melting is observed for NP diameters up to 30 nm. For larger sizes, the efficiency of the process decreases and the molten region no longer reaches the core of the NP such that it remains solid. Thus, NPs are only partially molten upon irradiation. Furthermore, The thickness of the liquid layer is observed to decrease with the NP size.

In a companion paper published in 2012 Rizza *et al.* give a rational description of the ion-shaping mechanism [69]. First, a matrix-like diagram is built to correlate the final ion-beam-shaped morphology to both the initial NP size and the irradiation fluence, figure 10. In a second step, 3DTS simulations are used to obtain the dependence of the vaporized/molten fraction on the initial NP size, figure 12c). Finally, the direct comparison between experimental results and simulations permits to obtain a precise characterization of four deformation regimes.

- Completely vaporized NPs (0-10 nm) remain spherical in shape upon irradiation.
- Completely molten NPs (10-35 nm) transform into nanorods and subsequently, for increasing fluence, into nanowires.
- Partially molten NPs (35-70 nm) transform into facetted NPs and subsequently, for increasing fluence, into nanowires.
- Very large NPs (>70 nm) do not melt and do not deform, or their deformation rate is very low.

The extension of each regime depends on both the NP composition and the irradiation parameters, such as the ion energy and the irradiation fluence.

Although, the papers of Dufour *et al* [87] and Rizza *et al* [69] represent a step forwards towards the comprehension of the ion-shaping mechanism, the fundamentals of the deformation process remain to be elucidated.

In this sense, an important insight has come from the work of Leino *et al.* published in 2014 [85]. Here, molecular dynamics simulations were used in conjunction with the inelastic (2D) thermal-spike model to include the electronic excitations of the swift heavy ions. The simulations predict that the elongation is caused by a dynamic crystal-liquid-crystal phase transition of the metal particle accompanied by a flow of the liquid phase into an under-dense track core in the silica matrix. The latter has been experimentally observed by Kluth *et al.* [89]. NPs whose minor axis is smaller than the diameter of the under-dense track expand more isotropically and are not elongated, which leads to a saturation of the width of the crystals. The results imply that both the formation of an under-dense track in the silica matrix and the shape-conserving crystallization of the molten cluster between ion impacts are necessary prerequisites to explain the elongation of metallic NPs by swift heavy-ion impacts.

6. Open problems, perspectives and challenges for the future

In this short historical overview, the evolution of the theoretical comprehension of two discoveries, the ion-hammering and the ion-shaping processes, has been traced from the first tentative steps to the state-of-the art description of the phenomenon. I have tried to show how this cognitive process is not linear and that back and forth oscillations are a necessary part of the maturation process. Nowadays, both phenomena have become versatile tools to shape the matter at the nanoscale. Indeed, both can be applied in nanofabrication methods allowing to control both the shape and the spatial orientation of dielectric nanocolloids and metallic NPs embedded within a dielectric matrix. However, they suffer from two main limitations: the radiation effect is size-dependent and the energy delivered by the impinging ions (either in the nuclear or electronic stopping regimes) scales with the length of the ion path. Thus, for more advanced applications a better control of the initial conditions is necessary. Those conditions include the spatial, size and depth distributions of the NPs. A possible issue is to replace the commonly used preparation technique, i.e. the ion-beam synthesis, by a preparation method based on colloidal chemistry or electron beam lithography. In this way a model system can be obtained that consists in a pattern of spatially mono-dispersed NPs confined to a unique plane. This model system can be used to obtain NP configurations that are barely realizable with the conventional bottom-up (i.e. the colloidal chemistry) or top-down (i.e. the electron beam lithography) nanofabrication techniques.

To conclude this review we discuss some possible applications:

The ion-hammering phenomenon is theoretically well described by the EFTA model [32]. The scope of possible practical applications remains however limited. The combination of techniques based on chemical synthesis and ion irradiation renders it possible to fabricate a new class of colloidal ellipsoids of continuously variable shapes with potential applications as photonic band-gap crystals [45], removable templates for the fabrication of macro-porous materials or masks for lithographic patterning [90].

Although the ion-hammering phenomenon is still waiting for a coherent description, the potential applications are instead very attractive. This is in part due to the plasmonic properties of the the ion-shaped NPs. As a spherical NP can be ion-shaped into the NW whose length can be as long as 400-500 nm, the corresponding plasmon resonances can range from the visible to the near-mid infrared. Thus, a broadband region of the optical spectrum can be explored. Hereafter two promising applications are shortly described:

A plasmon-based biosensor is based on the fact that, when a bio-molecular binding event occurs close to the surface of a noble metal nanostructure, the refractive index of the immediate environment surrounding the nanostructure becomes modified [91, 92]. A possible scheme for a sensor that would be able to detect low concentrations of molecules could be based on the idea that the tips of the ion-shaped NPs are situated either slightly below sample surface or touching it. This way, bio-molecular interactions at the surface of the nano-structures may lead directly to changes in the local refractive index; these changes can than be monitored via the wavelength shift of the LSPR peak.

Another potential application is in the field of plasmonic meta-materials [93–95]. Here tailor made nanocomposites - combinations of metallic and dielectric materials- can be designed such as to obtain optical properties that do not occr in nature. Hyperbolic meta-materials are a sub-class of plasmonic meta-materials displaying hyperbolic (or indefinite) dispersion [96]. This dispersion occurs when one of the principal components of the electric effective tensor has the opposite sign of the other two principal components. Such anisotropic structured materials exhibit distinctive properties, including a strong enhancement of spontaneous emission, a diverging density of states, a negative refraction index and enhanced super-lensing effects. Arrays of spatially oriented ion-shaped NPs may display such a hyperbolic behavior.

References

[1] Young D A *Nature* 1958 **182** 375
[2] Silk E C H and Barnes R S 1959 *Phil. Mag.* **4** 970
[3] Fleischer R L, Price P B and Walker R M 1965 *J. Appl. Phys.* **36** 3645
[4] Lifshitz I M, Kaganov M I and Tanatarov L V 1960 *J. Nucl. Energy* A **12** 69
[5] Klaumünzer S and Schumacher G 1983 *Phys. Rev. Lett.* **51** 1987
[6] Leteurtre J and Quéré Y 1972 *Irradiation Effects in fissile materials* (Defects in Crystalline Solids Vol. 6), ed S Amelinckx, R Gevers and J Nihoul (North Holland, Amsterdam) pp 25-47
[7] Klaumünzer S, Hou M D and Schumacher G 1986 *Phys. Rev. Lett.* **57** 850
[8] Klaumünzer S, Li C L and Schumacher G 1987 *Appl. Phys. Lett.* **51** 97

[9] Hou M D, Klaumünzer S and Schumacher G 1990 *Phys. Rev.* B **41** 1144
[10] Audouard A, Balanzat E, Jousset J C, Lesueur D and Thomé L 1993 *J. Phys. Condens. Mat.* **5** 995
[11] Klaumünzer S 2008 in *Ion Beam Science: Solved and Unsolved Problems (Part I, II)* Med. Phys. **35** p 293
[12] Klaumünzer S, Li C L, Löffler S, Rammensee M, Schumacher G and Neitzert HCh 1989 *Rad. Eff. Def. Sol.* **108** 131
[13] Benyagoub A, Lffler S, Rammensee M and Klaumnzer S 1992 *Nucl. Instrum. Meth.* B **65** 228
[14] Klaumünzer S 1989 *Rad. Eff. Def. Sol.* **110** 79
[15] S.Klaumünzer 2000 *Plastic flow of amorphous materials during ion-bombardment* in *Multiscale Phenomena in Plasticity: From Experiments to Phenomenology, Modelling and Materials Engineering* edsJ. Lpinoux *et al* NATO Science Series **367** p 441-450
[16] Klaumünze S and Benyagoub A 1991 *Phys. Rev.* B **43** 7502
[17] Lesueur D and Dunlop A 1993 *Rad. Eff. Def. Sol.* **126** 163
[18] Snoeks E, Polman A, and Volkert C A 1994 *Appl. Phys. Lett.* **65** 2487
[19] Snoeks E, Weber T, Cacciato A and Polman A 1995 *J. Appl. Phys.* **78** 4723
[20] Klaumünzer S 2004 *Nucl. Instr. Meth.* B **225** 136
[21] Schiwietz G, Grande P, Skogvall B, Biersack J P, Khrbrck R, Sommer K and Schmoldt A 1992 *Phys. Rev. Lett.* **69** 628
[22] Xiao G, Schiwietz G, Grande P L, Schmoldt A, Grether M, Köhrbrück R, Stolterfoht N, Spieler A and Stettner U 1996 *Nucl. Instrum. Meth.* B **115** 215
[23] Xiao G, Schiwietz G, Grande P L, Stolterfoht N, Schmoldt A, Grether M, Köhrbrück R, Spieler A and Stettner U 1997 *Phys. Rev. Lett.* **79** 1821
[24] Schiwietz G, Roth M, Czerski K, Staufenbiel F, Rösler M and Grande P L 2003 *Nucl. Instrum. Meth.* B **209** 26
[25] Schiwietz G, Czerski K, Roth M, Staufenbiel F and Grande P L 2004 *Nucl. Instrum. Meth.* B **226** 683
[26] Staufenbiel F, Schiwietz G, Czerski K, Roth M and Grande P L 2005 *Nucl. Instrum. Meth.* B **230** 426
[27] Toulemonde M, Dufour C and Paumier E 1993 *Phys. Rev.* B **46** 1436
[28] Szenes G 1995 *Phys. Rev.* B **51** 8026
[29] Trinkaus H and Ryazanov A I 1995 *Phys. Rev. Lett.* **74** 5072
[30] Trinkaus H 1995 *J. Nucl. Mat.* **223** 196
[31] Trinkaus H 1996 *Nucl. Instrum. Meth.* B **107** 155
[32] Trinkaus H 1998 *Nucl. Instrum. Meth.* B **146** 204
[33] Eshelby J D, 1957 *Proc. R. Soc.* A **241** 376
[34] van Dillen T, Onck P, van der Giessen E and Polman A 2005 *Phys. Rev.* B **71** 24103
[35] Benyagoub A, Klaumünzer S and Toulemonde M 1998 *Nucl. Instr. Meth.* B **146** 449
[36] Rotaru C 2002 PhD thesis, University of Caen Basse-Normandie
[37] Feynman R 1960*There's plenty of room at the bottom*, Engineering and Science, Vol. 2**3-5**, Caltech p 22
[38] Drexler K E 1986 *Engines of Creation: The Coming Era of Nanotechnology*, Publisher Doubleday
[39] Hémon S, Dufour C, Gourbilleau F, Paumier E, Dooryhée E and Bégin-Colin S 1998 *Nucl. Instrum. Meth.* B **146** 443
[40] Berthelot A, Hémon S, Gourbilleau F, Dufour C, Dooryhée E and Paumier E 1998 *Nucl. Instrum. Meth.* B **146** 437
[41] Snoeks E, van Blaaderen A, van Kats C M, Brongersma M L, van Dillen T and Polman A 2000 *Adv. Mat.* **12** 1511
[42] Snoeks E, van Blaaderen A, van Dillen T, van Kats C M, Brongersma M L and Polman A 2001 *Nucl. Instrum. Meth.* B **178** 62
[43] van Dillen T, Snoeks E, Fukarek W, van Kats C M, Velikov K P, van Blaaderen A and Polman 2001 *Nucl. Instrum. Meth.* B **175** 350
[44] van Dillen T, Polman A, van Kats C M and van Blaaderen A *Appl. Phys. Lett.* 2003 **83** 4315
[45] van Dillen T, van Blaaderen A and Polman A 2004 *Materials Today* p. 40
[46] van Dillen T, van der Giessen E, Onck P R and Polman A 2006 *Phys. Rev.* B **74** 132103
[47] Cartz L, Karioris F G and Fournelle R A 1981 *Radiat. Eff.* **54** 57
[48] van Dillen T 2004 *Ion irradiation-induced anisotropic plastic deformation* PhD thesis, Utrecht University (www.erbium.nl)
[49] Klaumünzer S 2004 *Nucl. Instrum. Meth.* B **215** 345
[50] Rizza G, Ramjauny Y, Hayoun M, Perruchas S, Gacoin T, Kluth P, Ridgway M C 2011 *Nanotechnology* **22** 475302
[51] D'Orléans C, Stoquert J P, Estourns C, Cerruti C, Grob J J, Guille J L, Haas F, Muller D and Richard-Plouet M 2003 *Phys. Rev.* B **67** 220101
[52] D'Orléans C, Stoquert J P, Estournes C, Grob J J, Muller D, Guille J L, Richard-Plouet M, Cerruti C and Haas F 2004 *Nucl. Instrum. Meth.* B **216** 372
[53] D'Orléans C, Stoquert J P, Estournés C, Grob J J, Muller D, Guille J L, Richard-Plouet M, Cerruti C and Haas F 2004 *Nucl. Instrum. Meth.* B **216** 372
[54] Harkati Kerboua C, Lamarre J M, Chicoine M, Martinu L and Roorda S 2013 T*hin Solid Films* **527** 186
[55] Joseph B, Ghatak J, Lenka H P, Kuiri P K, Sahu G, Mishra N C and Mahapatra D P 2007 *Nucl. Instrum. Meth.* B **256** 659
[56] Oliver A, Reyes-Esqueda J A, Cheang-Wong J C, Roman-Velazquez C E, Crespo-Sosa A, Rodriguez-Fernandez L, Seman J A and Noguez C 2006 *Phys. Rev.* B **74** 245425

Journal of Physics: Conference Series **629** (2015) 012005 doi:10.1088/1742-6596/629/1/012005

[57] Rodriguez-Iglesias V, Pena-Rodriguez O, Silva-Pereyra H G, Rodriguez-Fernandez L, Cheang-Wong J C, Crespo-Sosa A, Reyes-Esqueda J A and Oliver A 2010 *Opt. Lett.* **35** 703

[58] Yang Y, Zhang C, Song Y, Gou J, Zhang L, Meng Y, Zhang H and Ma Y 2013 *Nucl. Instrum. Meth.* B **308** 24

[59] Shirai M, Tsumori K, Kutsuwada M, Yasuda K and Matsumura S 2009 *Nucl. Instrum. Meth.* B **267** 1787

[60] Pivin J C, Singh F, Angelov O and Vincent L 2009 *J. Phys.* D **42** 025005

[61] Kluth P, Giulian R, Sprouster D J, Schnohr C S, Byrne A P, Cookson D J and Ridgway M C 2009 *J. Appl. Phys.* **94** 113107

[62] Dawi E A, Rizza G, Mink M P, Vredenberg A M and Habraken F 2009 *J. Appl. Phys.* **105** 074305

[63] Awazu K, Wang X, Fujimaki M, Tominaga J, Aiba H, Ohki Y and Komatsubara T 2008 *Phys. Rev.* B **78** 054102

[64] Ridgway M C, Kluth P, Giulian R, Sprouster D J, Araujo L L, Schnohr C S, Llewellyn D J, Byrne A P, Foran G J and Cookson D J 2009 *Nucl. Instrum. Meth.* B **267** 931

[65] Avasthi D K, Mishra Y K, Singh Fand Stoquert J P 2010 *Nucl. Instrum. Meth.* B **268** 3027

[66] Awazu K, Wanga X, Fujimaki M, Tominaga J, Fujii S, Aiba H, Ohki Y and Komatsubara T 2009 *Nucl. Instrum. Meth.* B **267** 941

[67] Awazu K, Wang X, Komatsubara T, Watanabe J, Matsumoto Y, Warisawa S and Ishihara S 2009 *Nanotechnology* **20** 325303

[68] Rizza G, Dawi E A, Vredenberg A M and Monnet I 2009 *Appl. Phys. Lett.* **95** 043105

[69] Rizza G, Coulon P E, Khomenkov V, Dufour C, Monnet I, Toulemonde M, Perruchas S, Gacoin T, Mailly D, Lafosse X, Ulysse C and Dawi E A 2012 *Phys. Rev.* B **86** 035450

[70] Rizza G, Attouchi F, Coulon P E, Perruchas S, Gacoin T, Monnet I and Largeau L, 2011 *Nanotechnology* **22** 175305

[71] Giulian R, Kluth P, Araujo L L, Sprouster D J, Byrne A P, Cookson D J and Ridgway M C 2008 *Phys. Rev.* B **78** 25413

[72] Sprouster D Jand Ridgway M C 2012 *Appl. Sci.* **2** 396

[73] Kluth P, Giulian R, Sprouster D J, Schnohr C S, Byrne A P, Cookson D J and Ridgway M C 2009 *Appl. Phys. Lett.* **94** 113107

[74] Ridgway M C, Giulian R, Sprouster D J, Kluth P, Araujo L L, Llewellyn D L, Byrne A P, Kremer F, Fichtner P F P , Rizza G, Amekura H and Toulemonde M 2011 *Phys. Rev. Lett.* **106** 095505

[75] Dawi E A, Vredenberg A M, Rizza G and Toulemonde M 2011 *Nanotechnology* **22** 215607

[76] Amekura H, Ishikawa N, Okubo N, Ridgway M C, Giulian R, Mitsuishi K, Nakayama Y, Buchal Ch, Mantl S and Kishimoto N 2011 *Phys. Rev.* B **83** 205401

[77] Schmidt B, Mueklich A, Roetzsch L and Heinig H-K 2007 *Nucl. Instrum. Meth.* B **257** 3

[78] Schmidt B, Heinig K H, Mueklich A and Akhmadaliev C 2009 *Nucl. Instrum. Meth.* B **267** 1345

[79] Araujo L L, Giulian R, Sprouster D J, Schnohr C S, Llewellyn D J, Johannessen B, Byrne A P and Ridgway M C 2012 *Phys. Rev.* B **85**

[80] Roorda S, van Dillen T, Polman A, Graf A, Vredenberg A M, van Blaaderen A and Kooi B 2004 *Adv. Mater.* **16** 235

[81] Penninkhof J J, van Dillen T, Polman A, Graf C and van Blaaderen A, 2005 *Adv. Mater.* **17** 1484

[82] Penninkhof J J, van Dillen T, Graf C, van Blaaderen A, Vredenberg A M and Polman A 2006 *Nucl. Instrum. Meth.* B **242** 523

[83] Penninkhof J J, Moroz A, van Blaaderen A and Polman A 2008*J. Phys. Chem.* C **112** 4146

[84] Penninkhof J J, Sweatlock L A, Moroz A, Atwater H A, van Blaaderen A and Polman A 2008 *J. Appl. Phys.* **103** 123105

[85] Leino A, Pakarinen O H, Djurabekova F, Nordlund K, Daraszewicz S, Kluth P and Ridgway M C 2014 *Mater. Res. Lett.* **2** 37

[86] Klaumünzer S 2006 *Nucl. Instrum. Meth.* B **244** 1

[87] Dufour C, Khomenkov V, Rizza G and Toulemonde M 2012 *J. Phys.* D **45** 065302

[88] Daraszewicz S L and Duffy M D 2011 *Nucl. Instrum. Meth.* B **269** 1646

[89] Kluth P, Schnohr C S, Pakarinen O H, Djurabekova F, Sprouster D J, Giulian R, Ridgway M C, Byrne A P, Trautmann C, Cookson D J, Nordlund K and Toulemonde M 2008 *Phys. Rev. Lett.* **101** 175503.

[90] Vossen D L J, Fific D, Penninkhof J, van Dillen T, Polman A and van Blaaderen A 2005 *Nano Lett.* **5** 1175

[91] Anker J N, Hall W P, Lyandres O, Shah N C, Zhao J and van Duyne R P 2008 *Nature Mat.* **7** 442

[92] Stewart M E, Anderton C R, Thompson L B, Maria J, Gray S K, Rogers J A and Nuzzo R G 2008 *Chem. Rev.* **108** 494

[93] Pendry J B, Schurig D and Smith D R 2007 *Science* **312** 1780

[94] Capolino F *Metamaterials Handbook 1: Theory and Phenomena of Metamaterials* , CRC Press, Boca Raton, CA

[95] Boltasseva A and Atwater H A2011 *Science* **331** 290

[96] Poddubny A, Iorsh I, Belov P and Kivshar Y 2013 *Nature Photonics* **7** 948

Journal of Physics: Conference Series **629** (2015) 012006 doi:10.1088/1742-6596/629/1/012006

Spectroscopic study of chemical modifications induced by swift heavy ions on polymers: the contribution of the CIRIL Platform and the CIMAP Laboratory

Yvette Ngono-Ravache[1]

[1] Centre de Recherche sur les Ions, les Matériaux et la Photonique, UMR6252
CEA/CNRS /ENSICAEN/Université de Caen-Basse Normandie,
CIMAP-CIRIL-Ganil, BP 5133, 14070 Caen Cedex 05, France

E-mail: ngono@ganil.fr

Abstract. This paper gathers results obtained on the chemical ageing of polymers, at the CIRIL platform, using Swift heavy ions (SHI) from the GANIL accelerator. Swift heavy ions induce high values of electronic stopping power or LET (Linear Energy Transfer) and deposit their energy in the polymer through electronic processes, in a few nanometer size cylinder centered on the ion path. This results in huge local doses and dose rates. Both defects created in the polymer chain and gas release were quantified using spectroscopic methods (FTIR and Residual gas analysis (RGA)). Defects created in polymers submitted to SHI can be separated in two main series: defects common to all ionizing radiations and defects specific to SHI. A common trend of the evolution of these defects, under inert environment, is the following: 1) for the first group of defects, in most of the polyolefins, there is a limited (if inexistent) effect of LET on the radiation chemical yield of creation at low doses. Among defects of this first series, the behavior of vinyl groups is particular, 2) LET effect on SHI specific defects (triple bonds and cumulenes) is tremendous. Triple bonds (alkynes, alkyl or aryl cyanates) are created after a LET threshold value, depending on the polymer chemical structure. The dose effect on macromolecular defects, under inert environment, is also presented. The study of the LET effect on gas release, in various polyolefines, gives an insight on the mechanism of bond cleavage in presence of high ionization and excitation densities. Finally, few results on radiation-induced oxidation are presented. Compared to low-ionizing radiations, oxidation is reduced and unsaturated bonds are created under SHI.

1. Introduction

Since the industrial development of organic synthetic polymers in the mid 20th century, the question of their behavior, first under weathering (UV irradiation) and then under ionizing radiation has been considered.

The effect of ionizing radiations on polymers can be either beneficial or detrimental. Low-ionizing radiations such as electron-beam (e-beam) or γ-rays can be used to increase some of the mechanical properties of the polymer, through controlled cross-linking. As an example heat shrinkable polyethylene (PE) for goods wrapping are obtained by low dose e-beam irradiation. In the same way, wooden art craft or antiquity can be rehabilitated through γ-curing of polyester resins. Besides, some of polymer-based medical disposal are sterilized using either γ-rays or e-beam at low doses. Unfortunately there are situations were irradiation is detrimental for the polymers; that is the case for insulation sheath of electrical cables in nuclear plants or for polymers composing glove boxes used during nuclear fuel handling and analyses processes. In the latter case, polymers are submitted concomitantly to α, β and γ rays.

First studies on polymer ageing under ionizing radiation were performed using γ-rays and e-beam. Ion beam irradiation of polymers was belated. This started with the study of the behavior of photoresists used during semiconductor doping. The great development of ion beam irradiation of polymers is related first to the research of particle detectors for monitoring low fluences, and second for either commercial manufacture of filters or polymer micro- and nano-structuration.

Ionizing radiations interact with the polymers through electronic processes, resulting in the creation of excitation and ionization. These initial species lead, after a time scale of around 10^{-12} s, to the formation of chemical modifications (called hereafter defects) in the polymer and the subsequent mechanical and optical modifications.

Organic synthetic polymers are defined as the repetition of a chemical unit, of small mass, through covalent bonds. They are composed of a backbone and, in some cases, have pendant groups or side chains. Polymers are very sensitive to ionizing radiations, as compared to other materials such as ceramics or steel. Ionizing radiations induce bonds scissions leading to chemical defects. These defects can be created either in the polymer chain in the form of new chemical groups (macromolecular defects) or as small molecules leaving the polymer in the form of gas (gas release). Both macromolecular defects and gas release are function of the polymer chemical structure and the irradiation conditions (projectile nature, environment, dose, dose rate, temperature…).

Gamma-rays and e-beams deposit their energy in the material rather homogeneously. On the contrary, swift heavy ions (SHI) deposit their energy through electronic processes occurring close to the ion path: the latent track. Most of the primary excitations and ionizations occur in a track core of a few nanometers. High energy electrons, emitted during ion/polymers interactions, induce ionization/excitations at larger distances from the ion trajectory, in the so-called track halo. This halo is defined by the range of secondary electrons, which increases with the projectile velocity [1]. The huge amount of energy locally deposited by SHI within the track can induce specific damage processes, which involve complex molecular rearrangements and collective atoms motion.

The LET induced by SHI in polymers are very high and lead to high ionizing and excitation densities. The comprehension of processes triggered by these high ionization/excitation densities was one of the motivation of the research, on SHI/polymers interactions, performed at CIRIL, using the GANIL beam lines, during the last 30 years. Basic research, aiming to the comprehension of the mechanisms leading to defects formation, as well as more applied research was considered. For the latter, the LET domain of interest is that of actinide-emitted α particles. In terms of LET, actinide-emitted α particles are simulated by 11 to 13 MeV/A energy C-Ne ion beams. Both macromolecular defects and gas emission were studied and their evolutions as a function of, the polymer chemical structure, the irradiation environment (vacuum or oxygen), the irradiation temperature (8 K, 298 K), the dose, the dose rate and the LET were followed.

2. Experimental

One of the advantages of polymer irradiation studies performed at CIRIL is the use of *on-line* analyses. The analytical approaches are essentially Residual gas analysis (RGA), Fourier transformed infrared and UV-visible spectroscopies. Specific equipments enable to irradiate and analyze the material without removing it from the set-up. There are two main advantages for this procedure. The first advantage is to probe the same point of the sample, thus avoiding possible heterogeneity in the sample thickness. The second advantage is first to prevent the post-irradiation oxidation prior the analysis (as polymers are known to be sensitive to radiation-induced oxidation), and second to avoid annealing, prior to analysis, after low temperature irradiations.

A large range of ions and energies was used for the actual studies. Most of them from the medium energy line of GANIL.

Various polymers were used. The chemical structures of the repeating units of most of them are presented in figure 1.

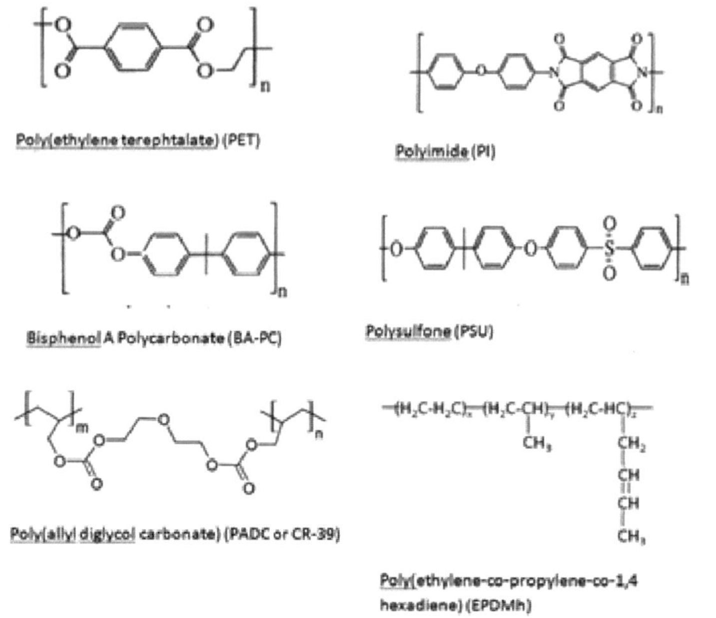

Figure 1. Chemical structures of the repeating units of some of the polymers presented in this paper.

Results can be discussed either as a function of the ion fluence or as a function of the dose deposited. Fluence (Φ in ion.cm^{-2}) and dose (D in Gy) are related through the LET (MeV.mg^{-1}.cm^{2}) according to the following equation: $D=1.6 \cdot 10^{-7} \cdot \Phi \cdot LET$.

Defects creation (or group destruction) is generally quantified using the radiation chemical yield of creation (or destruction), G. It represents the number of moles of defect created (or group destroyed) per unit dose and is expressed in mol.J^{-1}.

3. Results

In this section, for comprehension purposes, and since processes involved are quite different, results obtained during irradiation under vacuum will be separated from those concerning irradiations under oxidative environment.

3.1. SHI effect under vacuum or inert atmosphere

Under anoxic conditions (vacuum or He atmosphere), both macromolecular defects and gas release were monitored.

When a polymer is submitted to ionizing radiations, its FTIR spectra present two main features: 1) the decrease of some of the bands present in the spectrum of the virgin polymer and 2) the appearance and subsequent increase of new absorption bands. The first group of bands gives indications on the overall destruction of the initial polymer and the second group of bands is related to the creation of new defects in the polymer chain. Both features are shown in figure 2, which presents the evolution of the spectrum of a polyethylene (PE) film irradiated with a 6.6 MeV/A Sn ion beam [2].

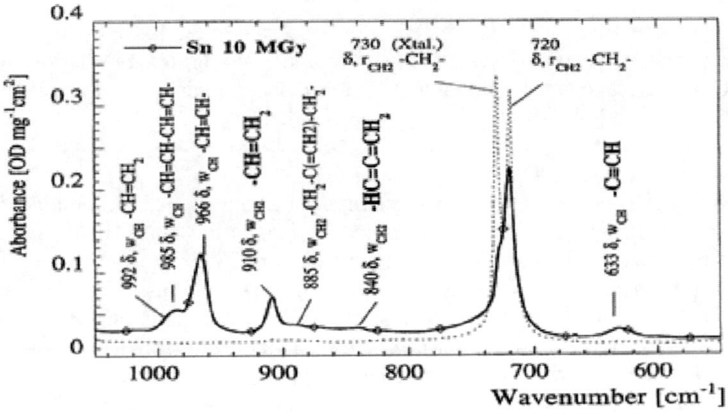

Figure 2. Infrared spectrum selected area, along with bands assignments, representing modifications induced in a semi-cristalline polyethylene (PE) thin film (37.8 μm). Dashed lines: virgin PE film. Solid lines: PE film irradiated at 10 MGy with a 6.6 MeV/A Sn beam. From reference [2].

Macromolecular defects created in the polymer chains irradiated under anoxic conditions are:

- **Double bonds**. They can be either isolated or conjugated alkenes. Isolated alkenes such as *trans*-vinylene, vinyl, and vinylidene are formed in almost all polymers under SHI. *Trans-trans*-dienes are also created but in a lesser extent. In polymers containing heteroatoms in the main chains, specific groups are created. As an example, in bisphenol A polycarbonate (BA-PC), carboxylic acids (along with phenols) are created after the carbonate group scission [3].
- **Triple bonds and cumulated double bonds**. In certain conditions, alkyne triple bonds are formed in polymers irradiated with swift heavy ions [4-7]. In Polyimide (PI), triple bonds formed are not only alkynes but also alkyl or aryl cyanates [8, 9]. Concerning cumulated double bonds, allenes and isocyanates are formed in polyolefines [7] and polyimide, respectively.
- **Cross-linking or chain scissions**. In this document, "chain scissions" stands for bond cleavages in the main chain (also called the backbone). The ratio between cross-linking and chain scissions depends on the chemical structure of the polymer. A tentative general trend to distinguish between polymers with prominent scissions or prominent cross-linking is the following. Chain scissions prevail in presence of a quaternary carbon atom, a carbonate or an ester group in the polymer backbone: as in BA-PC [3], in poly(allyl diglycol carbonate) [10] (PADC or CR-39) or in Poly(ethylene terephtalate) (PET) [8, 9, 11] . Defining cross-linking polymers is more difficult. In polymers containing at least one H atom on the carbon adjacent to a CH_2 group or in polymers containing alkenes, cross-linking is favored: as in polyethylene (PE) or in polystyrene (PS) [12]. The situation is rather complicated for polypropylene (PP).

Following the behavior of the absorption bands related to groups present in the virgin polymer, as a function of the irradiation conditions, gives information on the evolution of the crystalline ratio, the setting of chain scissions or the implication of these groups on the polymer ageing under irradiation. All these features are called hereafter polymer "overall destruction". For example, the destruction of ester or carbonate groups, when these are located in the backbone, is a good template for chain scissions since it leads to backbone length shortening. In polymer containing halogen atoms, chlorine or fluorine [13], the carbon-halogen bond is the weakest link.

For a thorough description of ion beam irradiations, the atomic number, the velocity and the fluence should be specified [1]. However, for quantification purposes, these parameters can suitably be reduced into two main parameters: the LET and the ion fluence or the dose deposited.

3.1.1. LET effect. Macromolecular defects created in polymer submitted to SHI can be classified into two series. The first series concerns defects created both under low-ionizing radiations and SHI. The second series concerns defects created solely under SHI or defects, such as vinyl groups in most of the polymers, which creation is highly increased with increasing LET.

3.1.1.1 Defects common to low and high ionizing radiations

In-chain unsaturated bonds. Under vacuum or inert environments, among macromolecular defects of the first series are C=C unsaturations, such as *trans*-vinylene, *trans-trans*-diene or vinylidene. In alkenes created in polymers containing halogen atoms, such as PVDF [2, 14], some of the olefinic hydrogen atoms can be substituted by halogen atoms. In simple polyolefins, LET effect on the alkenes of this series, at low doses, is reduced if not inexistent. In polymers containing a benzene ring, as PS, an influence of LET does exist and an increase in *trans*-vinylene and vinylidene radiation chemical yield at initial dose, G_0, is observed with increasing LET [5] .

Along with cross-linking, ***trans*-vinylenes** are one of the major defects created in PE. In this polymer, values of the *trans*-vinylene creation yield at initial dose, $G_0(tv)$, in a large domain of LET (0.44 keV/nm to 18.2 keV/nm) appear to be rather constant [2, 5, 7, 14]. A quite reduced variation in $G_0(tv)$ values was observed, when going from a 2.5 MeV e-beam to a ^{13}C 11 MeV/A ion beam [2]. In polymers containing native *trans*-vinylenes, the fate of these groups depends on the value of the initial concentration compared to the value at saturation, C_{sat} (see section 3.1.2.1), at a given LET. If the initial concentration is lower than C_{sat} then the group is further created. At the reverse, if the initial concentration is higher than C_{sat}, then the group is destroyed under irradiation [15]. At room temperature, in an ethylene-propylene diene monomer with 2.2% 1,4 hexadiene molar content ($EPDMh_{2.2}$), the *trans*-vinylene group is destroyed both under a 1 MeV e-beam and a 11 MeV/A ^{13}C ion beam [16]. At higher LET, when the dose increases, there is first destruction, followed by further creation. The dose at which the destruction is overwhelmed by the creation decreases with increasing LET [17].

Vinylidenes are created in polyolefins and their creation yield depends on the polymer chemical structure. In simple polyolefins and at low doses, their creation yields can be classified as followed: PP≥PB>PE. PB is the poly(butene-1). No tangible effect of the LET is observed for this defect in PP irradiated with ion beams starting from C to Kr.

***Trans-trans*-dienes** are conjugated structures made of C=C bonds. Their radiation chemical yield of creation is low compared to that of simple alkenes. The same trend, as for *trans*-vinylene, is observed in the evolution of *trans-trans*-diene as a function of the LET, in PE, PP, and PB [6].

For all of the alkenes of this first series, there is almost no LET effect on the radiation chemical yields at low doses, in most of aliphatic polyolefins. To the contrary, values of the concentration at saturation clearly increase when going from low-ionizing radiations to ion beams. Moreover, the saturation occurs at a lower dose for e-beam irradiations than for SHI. In PE, *trans*-vinylene concentration at saturation is multiplied by a factor of 2 when going from a 1 MeV e-beam ($1.84\ 10^{-4}$ eV/nm) to a 4.84 MeV/A ^{16}O beam (0.51 eV/nm); at the same dose rate. The same evolution is observed in poly(vinylidene fluoride), PVDF [2]. For both polymers, when increasing the LET, under SHI, from 0.51 to 8.52 eV/nm, the *trans*-vinylene concentration at saturation is almost constant. Since the saturation corresponds to a balance between creation and destruction, and since there is no LET effect on creation yields of at low doses, then the highest concentrations at saturation at high LET are assigned to reduced destruction efficiency.

Chain scissions and cross-linking. Direct analysis of both cross-linking and chain scissions by FTIR spectroscopy is difficult. Nevertheless, an indication of chain scissions can be obtained through the monitoring of the creation of vinyl groups (in polyolefins with no side chains) and triple bonds, the destruction of the carbonate or ester groups when present in the backbone (such as in BA-PC for example), or gas release. Since all of these features increases with increasing LET (sections 3.1.1.2 and 3.1.1.3), one can conclude on an increase in chain scissions with increasing LET.

The LET effect on cross-linking was studied in polystyrene irradiated either with a 2.5 MeV/A proton beam or a 9.8 MeV/A carbon beam, under vacuum and at low doses, leading to LET values of 0.02 and 0.18 keV/nm, respectively. This study showed first, that the partial yields for multiple cross-links increase with the LET increase and second, that a gel fraction can be created directly in the C ion track [12].

<u>*Polymer overall degradation.*</u> Following the evolution of groups present in the virgin polymer gives information on the polymer "overall degradation" and on the presence of weak groups or bonds in the polymer chain. For instance, in BA-PC irradiated with various ion beams, giving different LET values, the carbonate group is shown to be the most fragile group [3]. In PET, the ester group is the most fragile. In both polymers, the cleavage of these fragile groups leads to the creation of carboxylic acids, carbonyls other than acids or esters, alcohols or alkenes. In BA-PC, the effect of LET on the polymer "overall destruction" is huge: the destruction yield of carbonate groups increases with increasing LET. The most striking effect, which has also been shown in PS and in ethylene/styrene copolymers, is the loss of the benzene ring stability [3, 4, 5] at high LET. This reduced stability is due to an increase in the benzene ring opening with increasing LET; as shown by acetylene emission or the decrease of the FTIR bands of the benzene ring. This huge destruction of benzene ring corresponds to a progressive change in the radiolysis mechanism. As a consequence, because of their huge destruction yield, benzene rings can no longer be used to stabilize efficiently aliphatic polymers at very high LET.

In semi-crystalline polymers, the crystalline ratio decreases: amorphization occurs in the polymer. This amorphization was shown to be poorly sensitive to LET, in PE and PVDF [2, 14].

3.1.1.2 Defects specific to high ionizing radiations or related. In polymers devoid of native ***vinyl*** groups, these defects are created under e-beam or γ-rays but in very low concentrations. Under SHI, in polyolefins with no side chains, both their radiation chemical yield at low doses and their concentration at saturation increase sharply [18]. The vinyl group creation yield, in olefins homopolymers or copolymers such as PE, PP, or EPDMh, increases drastically first when going from e-beam to C ion beam, and second when increasing the LET from C to Pb ion beam [2, 14, 19]. In most of the polymers in which the formation of a vinyl group comes along with main chain bonds cleavage, an increase in the creation of vinyl groups with increasing LET gives an indirect evidence of the increase in chain scissions with increasing LET.

Most of ***alkynes*** studied are monosubtituted acetylenes. In PE submitted to a 1 MeV e-beam, at doses as high as 105 MGy, independently of the dose rate, alkynes are not created [14]. In the same polymer, submitted to ^{36}S 11 MeV/A ion beam, alkynes are created at doses as low as 1 MGy. Therefore, for alkyne to be created, a huge energy deposition in a limited time scale ($\sim 10^{-15}$ s) is mandatory. In most polymers, the simultaneous breakage of multiple bonds and the subsequent reorganization are required for alkyne to be created. In polymers containing benzene rings in their structure, alkyne creation originates from the ring destruction [5].

Under SHI, alkyne creation, since low doses, is only effective after a LET threshold [8, 9, 11], which depends on the polymer chemical structure. The LET threshold value is higher than 6.6 $MeV.mg^{-1}.cm^2$, in PE, PP or PB [7]. In these polymers, above the LET threshold, the radiation chemical yield of creation of monosubstituted acetylene is function of the side chain length: it decreases when going from PE to PB. This behavior is LET dependant [7]. The present behavior can be only understood if one considers, at high LET, the increase of highly unsaturated hydrocarbon gases created from the side chain departure [20, 21]. Below the LET threshold, alkynes are created after a threshold of dose of around 10 MGy, independently of the side group length [7]. This threshold of dose was shown to decrease with increasing LET in α-PVDF [22].

LET effect on cumulene is similar to the one observed on alkynes: a LET Threshold is needed for them to be created and, above the threshold, their creation yield increases with the LET.

In polyimide, triple bonds and cumulenes created include a nitrogen atom to form respectively cyanates and isocyanates [8].

Both the LET threshold for alkyne creation and the threshold of dose, when the ion LET is below the LET threshold of creation, can be modified by modifying the polymer chemical structure. Inserting double bonds in the polymer reduces the LET threshold. This phenomenon is assigned to the increase in the electronic density in the environment of unsaturated bonds and the set-up of important energy transfers from the saturated moiety to the C=C bond [7].

3.1.1.3 Gas release. In simple polyolefins with no heteroatom, one of the most important gases emitted is hydrogen (H_2). This gas is formed from H°, resulting from C-H homolytic scissions, which reacts within the polymer. Apart from H_2, hydrocarbon gases are created in polyolefins under SHI. Probing hydrocarbon gases is an indirect way to access to chain scissions. In polymers with no side chains or pendant groups, the emission of hydrocarbons is related to bond cleavage in the polymer backbone.

At low LET, hydrocarbon gases are fingerprints of the polymer chemical structure. In polyolefins with side chains, the complementary gases are representative of the chemical structure of the side chain (CH_4 for PP, C_2H_x for PB…). Based on this behavior, one can infer that bond scissions occur at the side chain, on the tertiary carbon. When the LET is increased, the degree of unsaturation of hydrocarbon gases increases. At very high LET, the chemical structure signature is lost and acetylene becomes the preponderant gas among hydrocarbons [20] (table 1).

Another effect of the LET on gas emission is the increase of the radiation chemical yield of creation of hydrocarbon gases in all the polymers studied (PE, PP, PB, PS and EPDMh), with increasing LET. At LET as high as 20 MeV.mg^{-1}.cm^2, the radiation chemical yields of hydrocarbon gas emission in PS and PE are almost equal; indicating the lost in stability of styrene groups [5, 20, 21, 23].

Table 1. Evolution of hydrocarbon gaseous molecules release in PE, PP and PB irradiated at different LET, at a dose of 2 MGy, under vacuum and at room temperature (from reference [20]).

	^{13}C			^{14}N			^{58}Ni		
LET (MeV. mg^{-1}.cm^2)	1.8			6.8			39.7		
Polymer	**PE**	**PP**	**PB**	**PE**	**PP**	**PB**	**PE**	**PP**	**PB**
Total gas G value (10^{-7} mol/J)	0.03	0.16	0.4	0.2	0.5	1.4	1.7	3	3.5
Number of C atoms in the molecule (%)									
1	10.8	87.4	9.0	4.2	38	7	8	30	10
2	42	8	73	84.4	42	76.7	81.5	51.3	71
3	7.7	4.4	4	2.0	13.3	0.25	4	13.3	8
4	39.8	0	13.7	9.4	7	16	6.6	5.7	10.5
Percentage of alkanes, alkenes and alkynes									
Alkanes	72	87	57	15	49	28	13	34	24
Alkenes	11	6	29	18	15	29	19	21	21
Alkynes	17	7	13	67	37	43	68	45	55

In polymers containing ester groups (PET) or carbonate groups (BA-PC, CR-39), carbon oxides are among the most important gases emitted. These gases are created through the scission of the backbone on the ester or the carbonate site, depending on the polymer structure. Since the destruction of these groups was shown to increase with increasing LET [3], one can assume that the resulting gases would increase as the LET increases. In an aromatic polyimide irradiated at cryogenic temperatures [24], the major gaseous product is CO. Besides this molecule, CO_2 and hydrocarbon molecules (C_2H_x) are emitted. All of these gaseous products are indications of chain scissions.

In polymers containing benzene rings such as PI, PS, PET or BA-PC, benzene derivatives are present in the radiation-induced emitted gases at low LET. At higher LET, their ratio in the gas mixture is reduced at the benefice of acetylene coming from benzene ring opening [5, 23].

3.1.2. Dose effect

3.1.2.1 Macromolecular defects. Under SHI, the influence of the dose varies with 1) the defect considered, 2) the irradiation temperature and 3) the polymer chemical structure. Under inert atmosphere and at room temperature, for most of the macromolecular defects created in polymers, there is a tendency towards saturation: the concentration increases with increasing dose then levels off, at the saturation concentration (C_{sat}) [2-5, 13, 16, 18]. This means their radiation chemical yield of creation, or destruction, decreases with increasing dose. Under ionizing radiations, creation and destruction happen simultaneously. Therefore, at saturation, destruction is strictly compensated by creation. This saturation can be ascribed to track overlapping at high doses. An example of this evolution is shown on figure 3, for alkyne creation.

At a given LET and in a given polymer, at room temperature, when a chemical group that can also be created under irradiation is present in the pristine polymer, its evolution as a function of dose is more complex. When the concentration of this group in the pristine polymer is higher than its concentration at saturation, when radiation-induced, then it is destroyed under irradiation. To the contrary, when its concentration in the pristine polymer is lower than the concentration at saturation, when radiation-induced, it is created under irradiation. This was particularly studied in PE containing chemically inserted C=C of the *trans*-vinylene type [15].

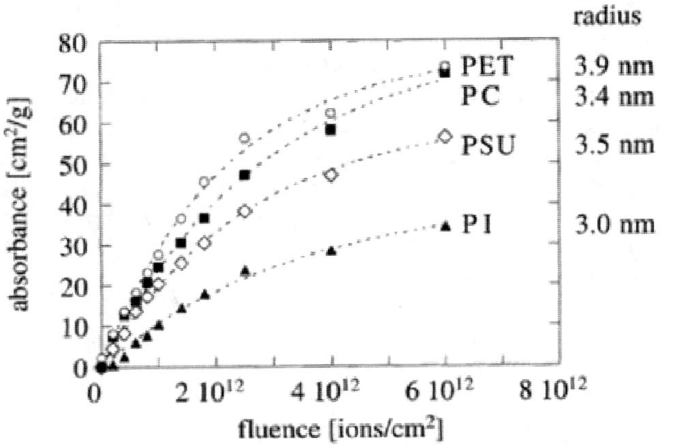

Figure 3. Normalized absorbance of alkyne band (ν_{CH} stretching vibration in RC≡CH, 3295 cm^{-1}) and related creation radii, in various polymers irradiated with Kr ions (720 MeV) under vacuum (from reference [9]). Creation radii values are extracted from an exponential fit, as assumed from the saturated track model.

One of the initial defect precursors in polyolefins is the C-H bond homolytic scission. This leads to the emission of H_2 along with the creation of unsaturated bonds of the alkene type. In polymers containing chemical functions such as esters and carbonates, or carbon-halogen bonds, the unsaturated bonds are created from the cleavage of these specific features. At limited doses, the prevalence of a given defect type is function of the polymer chemical structure. But, if one increases the irradiation dose up to very high values, all the polymers tend to merge into a carbon like structure made of aromatic structures containing aliphatic moieties [25].This structure is obtained by the progressive change in the polymeric structure with increasing dose, through the increase in the number of conjugated unsaturated bonds, as demonstrated by Infrared spectroscopy [2].

In semi-crystalline polymers, the crystalline ratio decreases with increasing dose and can almost disappear at very high doses.

The radius, R, in which a given defect is formed, can be calculated assuming a saturated track model. According to this model, defects are located exclusively in the latent track, which is continuous. The leveling off of the defects concentration, with the fluence, results from overlapping of latent tracks. From this, the evolution of the concentration (C) with the fluence (Φ) can be fitted by an exponential law following $C=C_{sat}(1-\exp(-\pi R^2\Phi))$. Based on this model, spatial distribution of defects within the track was calculated in PVDF [22] (figure 4) as well as in PE [6].

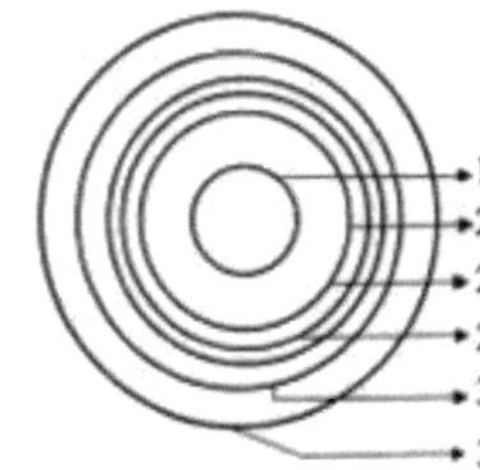

Figure 4. Spatial distribution of defects within the latent track in α-PVDF irradiated with a 10.6 MeV/A Ar beam, at room temperature (from reference [22]).

The value of the creation radius, R, is as high as the defect creation is easy. For Both PVDF and PE, whatever the defect considered, the creation radius increases with increasing LET [22, 6].

3.1.2.2 Gas release. As mentioned previously, the hydrogen molecule is the major gaseous molecule emitted from olefin homopolymers or their copolymers, in the absence of heteroatoms. In these polymers, the radiation chemical yield of creation of H_2 decreases with increasing dose and then levels off. This decrease has been studied more precisely for PE and EPDMh. It has been shown, experimentally, to be due to the set-up of energy transfers from the non-modified part of the polymer to defects previously created in the polymer chain [15].

In PE and PP, the emission yields of hydrocarbon gases increase with increasing dose. The total creation yield of hydrocarbon gases is multiplied by a factor of 1.5 in PE and 2 in PP, when increasing the irradiation dose from 2 MGy to 22 MGy, under ^{17}O (4.6 $MeV.mg^{-1}.cm^2$) beam. In the course of the increase of the creation yield of hydrocarbon gases, the ratio of unsaturated gases (alkenes and alkynes) over alkanes is increased. Consequently, chain scissions increase with increasing dose and this is accompanied by further dehydrogenation [23].

3.1.3 Influence of the irradiation temperature. The mechanisms underlying defect creation under SHI were studied using irradiation at very low temperatures (8 K-15 K), on different polymers or organic compounds. Despite their relatively simple chemical structures, it is difficult to evaluate mechanisms involved in polymers ageing, due to their chain length. Therefore, studying the effect of irradiation temperatures on polymer ageing can give valuable information on defect creation mechanisms. Studying polymers behavior at different irradiation temperatures or under annealing after irradiation at low temperatures has permitted to assess one of the most important questions concerning the formation of stable groups from radicals [19].

In PE, both ***trans*-vinylene** and ***cross-linking*** are formed from alkyl radicals. It was shown that, at temperatures where chains are frozen (8 K), only *trans*-vinylene groups are created. When the temperature rises close to the glass transition temperature, cross-linkings are formed in majority and part of the alkyl radicals leads to allyl radicals and to a lesser extent to *trans-trans*-diene [19].
In the same study, comparing radiation chemical yields at initial dose at 8 K and at room temperature showed that, half of the *trans*-vinylene is formed in the cage (immediate vicinity of the alkyl radical creation site) and half outside the cage. Therefore, only half of the *trans*-vinylene created at room temperature is diffusion-dependent [16, 18].

A similar phenomenon is proposed to explain the evolution of the *trans*-vinylene concentration in $EPDMh_{2.2}$ containing native *trans*-vinylene. In this polymer, the temperature effect on the C=C behavior is tremendous at low LET. At low temperatures (8 K) *trans*-vinylene concentration increases, with increasing dose, then saturates. To the contrary, at room temperature, *trans*-vinylene groups are consumed down to a saturation concentration. This consumption is assigned to C=C attack by migrating radicals [16].

Trans-trans*-dienes** are not created in PE irradiated at low temperature (8 K). Thus, these defects are not created in the cage. As a consequence, migration of reactive species is mandatory for *trans-trans*-diene to be created. In the same way, species migration was shown to be of great importance for ***allyl radical creation in PE. In the presence of native unsaturated bonds in the pristine polymer, the temperature effect on both *trans-trans*-diene and allyl radicals creation is very low [16].

Concerning ***vinyls*** and ***alkynes*** of the monosubstituted acetylene type, the influence of the irradiation temperature is clearly visible, in PE. When lowering the irradiation temperature from room temperature to 8 K, the alkyne creation yield, G(alkynes), increases. The ratio of G(alkynes), in PE [18], when going from RT to 8 K is 2.6. Consequently, although rapid, alkynes formation time scale is low enough to allow initial species migration. Thus, the creation time scale is higher than the diffusion time scale allowing the most mobile radicals, H°, to escape from their creation site.

Studying the behavior of simple molecule blends, at very low temperatures (11 K), enabled the comprehension of mechanisms in action during polymer irradiation with SHI. As stated above, aromatic molecules or benzene rings have been used since decades to stabilize aliphatic polymers under low-ionization irradiations, via energy transfers. Excitations and reactive species created in the aliphatic parts are transferred to the aromatic moiety. This stabilization effect still exists under SHI at low LET. Thus, despite de huge ionization/excitation densities induced by SHI, transfers exist and are efficient. However, the stabilization is lost at LET values higher than 7 $MeV.mg^{-1}.cm^2$; thanks to benzene ring opening. Furthermore, against the general knowledge, it has been shown (figure 5), using ethylene/styrene copolymers or amorphous solids of organic molecules, that the aliphatic moiety is protected at the expense of the aromatic moiety that is destroyed [26-28].

Transfers in polymers can happen either along a given chain (intrachain processes) or between two chains (interchain processes). In order to assess the importance of each of these processes, the behavior of cyclohexane/benzene amorphous solid mixtures, under SHI, was considered. It appears from this study that, contrary to what was considered in the past, excitations and charge transfers can also happen through the interchain route; and not solely through the intrachain route [26-28]. Besides, in ethylene/styrene copolymers with various styrene molar contents, increasing the irradiation temperature from 8 K to 298 K only showed a limited variation in the radiation chemical yields of defect creation. Therefore, the influence of reactive species, such as radicals, in the stabilization mechanism of aliphatic moiety by the aromatic one, under SHI, is reduced.

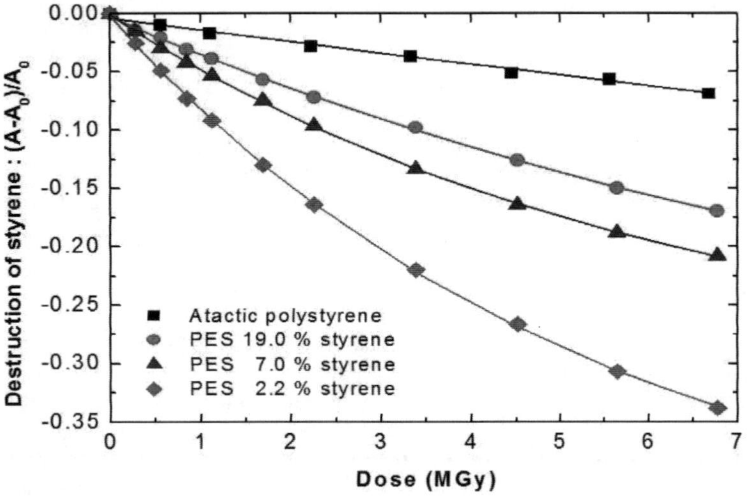

Figure 5. Evolution of the styrene moiety destruction ($(A-A_0)/A_0$) as a function of the dose in different Etylene/Styrene copolymers irradiated with ^{13}C ions (LET = 1.7 $MeV.mg^{-1}.cm^2$), under vacuum and at RT (from reference [28]). A_0 and A correspond to the absorbance before irradiation and at a given dose, respectively.

3.2. Influence of the irradiation environment: the radiation-induced oxidation

Studies on the combined effect of oxygen and SHI are less numerous and detailed than those concerning irradiations under anoxic conditions. In the following, $EPDMh_{2.2}$ is an ethylene-propylene-diene terpolymer based on 1,4 hexadiene (2.2% molar).

Under low-ionizing radiations, when oxygen is present in the irradiation environment, radicals created from C-H (or C-C) homolytic scissions, readily react with oxygen molecules giving peroxy radicals. According to the generally accepted chain reaction, involving alkyl and peroxy radicals, inactive products are created. These products are made of both oxidized and non-oxidized groups. Detailed studies on the radiation-induced oxidation of $EPDMh_{2.2}$, irradiated with various ion beams were handled [17, 29-31]. It appears from these studies that: first the oxidative ageing is effective under SHI and second, the oxygen uptake is lower under SHI. Oxygen uptake is multiplied by five when going from a 13 MeV/A ^{13}C beam irradiation to a 1 MeV e-beam irradiation.

In $EPDMh_{2.2}$ irradiated with a 13 MeV/A ^{13}C beam, oxidation occurs since very low oxygen pressures in the irradiation cell. The radiation chemical yield of oxygen uptake is function of the oxygen relative pressure in the irradiation vessel: it increases up to 10 mbar and then levels off [29].

Oxidized defects created under SHI are of the same type as those created under low ionizing radiations. However, their radiation chemical yields of creation are lower under SHI, as is the oxygen uptake [29,30]. This reduced oxidation effect can be due to 1) the huge energy deposit, corresponding to very high local dose rates, induced by SHI or 2) the use of higher macroscopic dose rates during irradiations with SHI or 3) the combination of both effects. The first hypothesis seems by far the most credible.

Possible oxidised defects are carbonyl groups (ketones, carboxylic acids, esters, peresters or peracids), ethers, alcohols or hydroperoxides. No LET effect was observed on total carbonyl concentration, in $EPDMh_{2.2}$ irradiated under Ne beams at LET ranging from 4.8 to 11.5 $MeV.mg^{-1}.cm^2$. Among carbonyl containing functions, thanks to curve fitting, ketones were shown to accumulate more than carboxylic acids; even if the ratio of ketones to carboxylic acids decreases with increasing dose [17].

Contrary to what is observed under low ionizing radiations, unsaturated groups are created under SHI in presence of oxygen. Creation of unsaturated groups under SHI can be accounted for the very high ionizing densities created along the ion track and the resulting enhanced alkyl radical recombinations, prior to reaction with oxygen molecule. Alkenes, created are of the same type as those created under inert environments. In $EPDMh_{2.2}$, irradiated at LET varying from 4.8 to 11.5 $MeV.mg^{-1}.cm^2$, in presence of oxygen, vinyl creation increases with increasing LET. Moreover, this LET effect is dose dependent [17]. In PVDF irradiated with a 5.3 MeV/A Kr beam, the creation of conjugated double bonds is enhanced under oxygen atmosphere [13]. Alkynes are created under oxidative environment, above a LET threshold and no effect of the irradiation environment was observed in their creation in PVDF films [13] or in $EPDMh_{2.2}$ films [17].

As a consequence of the oxygen uptake and the subsequent defects created, most of the polymers are degraded under oxidation. For those that are degrading under vacuum, the degradation rate is increased [8].

Under low-ionizing radiations, the oxygen uptake radiation chemical yield, $G(-O_2)$, in polymers evolves linearly with the reverse of the square root of the dose rate (I) following: $G(-O_2)=a+b/\sqrt{I}$. Under a 11 MeV/A ^{13}C beam, whatever the oxygen relative pressure, no dose rate effect was observed in the $G(-O_2)$ values in the $EPDMh_{2.2}$ at dose rates varying from 100 kGy/h to 1000 kGy/h [30]. The situation is different for some of the oxidized macromolecular defects [31]. Indeed, studies on the same EPDM hexadiene, irradiated with e-beam and ^{13}C 11 MeV/A beam under 200 mbar oxygen, showed that, for carbonyls or hydroxyls groups 1) the radiation chemical yields of creation are related to the dose rate square root by the linear relation presented upper, 2) the values of a and b are lower under ^{13}C beam irradiations as compared to e-beam. These lower values can be understood if one considers the higher local dose rate induced by SHI along the ion path.

4. Conclusion

An important effort was made to understand the behavior of polymers under SHI. Most of the studies were performed with, olefins homopolymers or their copolymers, PVDF, BA-PC, PS and Ethylene/styrene copolymers; under inert atmosphere. Studying the emission of radiation-induced hydrocarbon gases and their evolution as a function of the LET gives an insight on mechanisms involved on polymers ageing and their dependence with the density of excitations and ionizations.

Besides, varying the irradiation temperature is of great interest for the comprehension of mechanisms underlying the creation of stable defects.

Despite the huge energy deposition and the subsequent high ionization and excitation density induced by SHI, phenomena such as electronic excitation or reactive species transfers do exist. Besides, it has been shown that the aliphatic moiety stabilization occurs at the expense of the aromatic moiety.

The increase in the hydrocarbon gas emission as well as the increase in vinyl groups (in most of the polymers) and in alkyne groups creation with increasing LET gives a clear indication on the increase of chain scissions with increasing LET. The stability of polymers containing benzene rings, known and used under low-ionizing radiations, still exist under SHI at moderate LET. At the highest LET, this stability is lost due to the ring opening. Consequently, the radiation stabilization of aliphatic polymers by benzene ring insertion is no longer efficient at LET higher than 7 $MeV.mg^{-1}.cm^{2}$.

Few results were obtained under oxidative environment. Therefore great efforts should be oriented towards the comprehension of the influence of high LET on radiation-induced oxidation, especially the fate of non oxidized defects and hydrocarbon gas emission and their evolution either with irradiation dose, LET or dose rate.

References

[1] Bouffard S, Gervais B and Leroy C 1995 *Nucl. Instrum. Meth.* B **105** 1-4
[2] Balanzat E, Betz N and Bouffard S 1995 *Nucl. Instrum. Meth.* B **105** 46-54
[3] Dehaye F, Balanzat E, Ferain E and Legras R 2003 *Nucl. Instrum. Meth.* B **209** 103-112
[4] Balanzat E, Bouffard S, Dévy J and Gaté Chr 1996 *Nucl. Instrum. Meth.* B **116** 159-163
[5] Ferry M, Ngono-Ravache Y, Picq V and Balanzat E 2008 *J. Phys.Chem.* B **112** 10879-10889
[6] Gaté Ch *Polymères aliphatiques sous irradiation par des ions lourds* 1997 PhD thesis, Université de Caen Basse Normandie
[7] Ngono-Ravache Y, Corbin D, Gaté Chr, Mélot M and Balanzat E 2007 *J. Phys. Chem.* B **111** 2813-2819
[8] Steckenreiter T, Balanzat E, Fuess H and Trautmann C 1999 *J. Polym. Sci.* A **37** 4318-4329
[9] Steckenreiter T, Balanzat E, Fuess H and Trautmann C 1999 *Nucl. Instrum. Meth.* B **151** 161-168
[10] Yamauchi T, Barillon R, Balanzat E, Asuka T *et al* 2005 *Rad. Measur.* **40** 224-228
[11] Steckenreiter T, Balanzat E, Fuess H and Trautmann C 1997 *Nucl. Instrum. Meth.* B **131** 159-166
[12] Bouffard S, Balanzat E, Leroy C, Busnel J.P and Guevelou G 1997 *Nucl. Instrum. Meth.* B **131** 79-84
[13] Betz N, Le Moel A, Balanzat E, Lamotte J, Gallas J.P and Jaskierowicz G 1994 *J. Polym. Sci.* A. **32** 1493-1502
[14] Balanzat E, Bouffard S, Le Moel E and Betz N 1994 *Nucl. Instrum. Meth.* B **91** 140-145
[15] Ventura A *Polymères sous rayonnements ionisants : étude des transferts d'énergie vers les défauts d'irradiation* 2013 PhD thesis. Université de Caen Basse-Normandie
[16] Mélot M, Ngono-Ravache Y and Balanzat E 2003 *Nucl. Instrum. Meth.* B **208** 345-352
[17] Ngono-Ravache Y, Damaj Z *et al* 2015 *Polym. Degrad. Stabil.* **111** 89-101
[18] Mélot M, Ngono-Ravache Y, Balanzat E 2003 *Nucl. Instrum. Meth.* B **209** 205-211
[19] Mélot M *Matériaux organiques irradiés à très basse température et à différents pouvoirs d'arrêt : cas du polyéthylène et des molécules de cyclohexane isolées en matrice* 2003 PhD thesis, Université de Caen Basse Normandie
[20] Picq V, Ramillon J M and Balanzat E 1998 *Nucl. Instrum. Meth.* B **146** 496-503
[21] Picq V and Balanzat E 1999 *Nucl. Instrum. Meth.* B **151** 76-83
[22] Le Bouëdec A, Betz N, Esnouf S, Le Moël A 1999 *Nucl. Instrum. Meth.* B **151** 89-96
[23] Picq V *L'émission gazeuse des polymères aliphatiques sous irradiation : effet du pouvoir d'arrêt* 2000 PhD thesis, Université de Caen Basse-Normandie
[24] Severin D, Balanzat E, Ensinger W, and Trautmann C 2010 *J. App. Phys.* **108** 024901
[25] Reynaud C, Guillois O, Herlin-Boime N, J-N.Rouzaud *et al* 2001 *Spectrochim. Acta* A **57** 797-814
[26] Ferry M *Comportement des verres cyclohexane/benzène et des copolymères éthylène/styrène sous rayonnements ionisants:transferts d'énergie et d'espèces entre groupements aliphatiques et aromatiques* 2008 PhD thesis, Université de Caen Basse-Normandie

[27] Ferry M, Bessy E, Harris H *et al* 2013 *J. Phys. Chem.* B **117** 14497-14508
[28] Ferry M, Bessy E, Harris H *et al* 2012 *J. Phys. Chem.* B **116** 1772-1776
[29] Dély N, Ngono-Ravache Y, Ramillon J-M and Balanzat 2005 *Nucl. Instrum. Meth.* B **236** 145-152
[30] Dély N *Radio-oxydation d'un élastomère de type EPDM lors d'irradiations faiblement ou fortement ionisantes : mesure et modélisation de la consommation d'oxygène* 2005 PhD thesis, Université de Caen Basse-Normandie
[31] Corbin D *Etude de l'oxydation et de la tenue d'élastomères irradiés : conséquences sur l'intégrité des câbles électriques lors d'une situation accidentelle d'un réacteur à eau pressurisée* 2001 PhD thesis, Université de Caen Basse Normandie

Journal of Physics: Conference Series **629** (2015) 012007 doi:10.1088/1742-6596/629/1/012007

Swift heavy ion induced electron emission from solids

Hermann Rothard[1], Gaetano Lanzanò[2], Benoit Gervais[1], Enrico De Filippo[2], Michel Caron[3] and Michael Beuve[4]

[1] Centre de Recherche sur les Ions, les Matériaux et la Photonique (UMR6252), CEA/CNRS/ENSICAEN/Université de Caen-Basse Normandie UCBN, CIMAP-CIRIL-Ganil, BP 5133, 14070 Caen Cedex 05, France
[2] Istituto Nazionale di Fisica Nucleare INFN Sezione di Catania, Via S. Sofia 64, 95123 Catania, Italy
[3] CEA-GRAMAT, Route de Reilhac, BP800, F-46500 GRAMAT, France
[4] Institut de Physique Nucléaire de Lyon IPNL,UMR 5822, Université de Lyon, Université Lyon 1, CNRS/IN2P3, 4 rue E. Fermi, 69622 Villeurbanne Cedex, France

E-mail: hermann.rothard@ganil.fr

Abstract. We briefly summarize the results of numerous experiments performed at GANIL aimed at measuring electron yields and doubly differential yields (energy or velocity spectra at different ejection angles, angular distributions). These studies, supported by theoretical investigations and numerical simulations, contributed decisively to our understanding of the very first step in energy deposition in matter, i.e. ionization and subsequent electron transport through condensed matter. The emitted electron spectrum contains a rich variety of features including binary encounter electrons (BEE), convoy electrons (CE), Auger electrons (AE) and the low-energy peak of "secondary" electrons (SE).

1. Electron Ejection by swift heavy ions at CIRIL-GANIL

Radiation effects in condensed matter emerge from "primary ionization" (and excitation) of target atoms, which leads to liberation of electrons. Electrons play thus a key role in track formation and in the related damage creation. A detailed knowledge of the structure of ion tracks is a key issue for our understanding of radiation effects in condensed matter (track formation, radiolysis, RBE calculations for hadrontherapy, etc.), where doubly differential electron ejection cross sections (DDCS) are a key input parameter for numerical simulations. Experimentally, electron yields, i.e. the mean number of emitted electrons, have been studied. Differential cross sections are measured with either magnetic or electrostatic spectrometers. Furthermore, the time-of-flight technique was used. GANIL is particularly well suited for this method, since it delivers pulsed beams with good time resolution (< 1 ns). The spectrometer is placed at a certain observation angle θ. The ejection angle θ is usually measured with respect to the beam direction (θ = 0 deg.). Often, thin foils are used as targets, which allows in addition to "backward" emission (θ > 90 deg.) also to study electron ejection in "forward direction" (θ < 90 deg.). Figure 1, taken from a review on achievements and open questions in electron ejection by swift ions [1], shows schematically some of the mechanisms involved.

Fast electrons induced by swift ions are often referred to as "δ-rays" or "δ-electrons". The electrons kicked by the projectile may then undergo elastic or inelastic scattering during their travel through the medium ("electron transport"). The inelastic scattering liberates target electrons, thus creating in turn,

Journal of Physics: Conference Series **629** (2015) 012007 doi:10.1088/1742-6596/629/1/012007

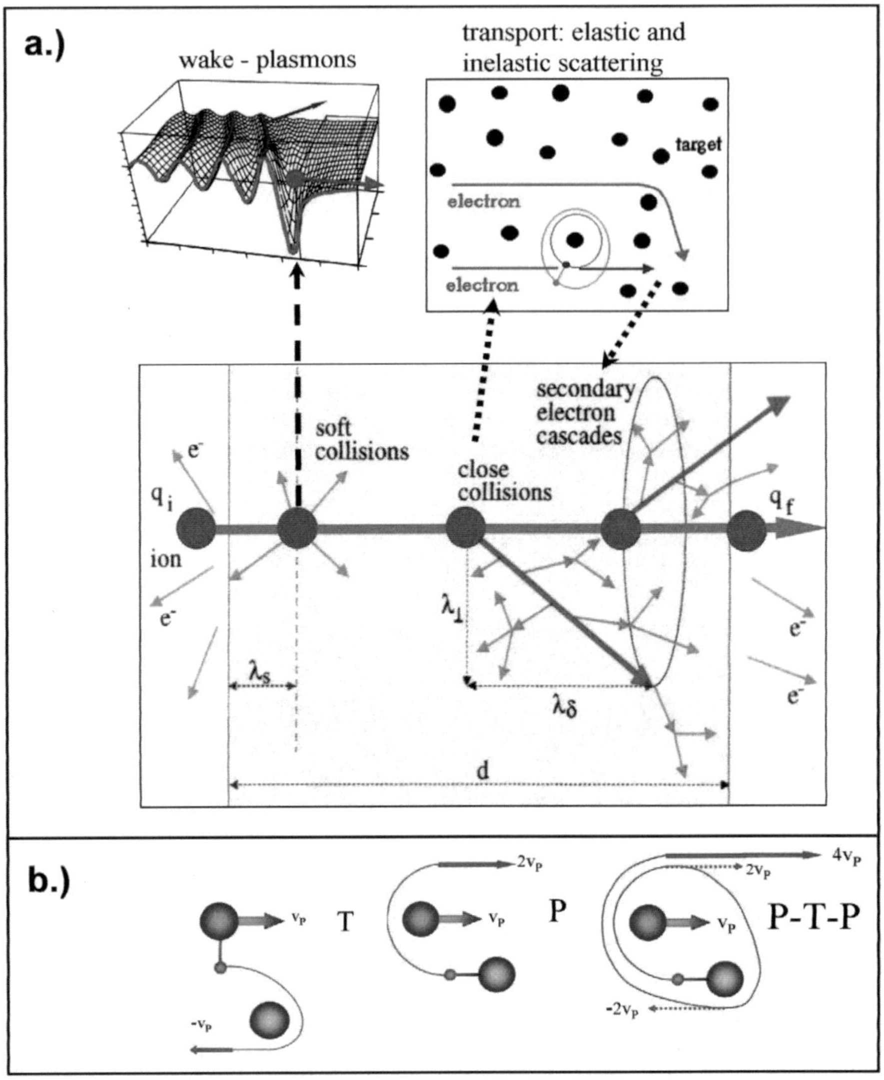

Figure 1. Schematic representation of mechanisms intervening in electron ejection from a thin solid foil of thickness d bombarded with an ion beam of incoming charge q_i and outgoing charge q_f. Primary ionization in close collisions leads to ejection of fast electrons from the target atoms ("binary encounter", denoted P, bottom part). Low energy electrons stem from ionization in distant collisions or collective effects (plasmon decay, wake, upper left part). On their way through the solid towards the surface, electrons suffer inelastic collisions without large angular deflections and angular scattering by the screened nuclei (upper right part). Also, an electron carrying projectile can be ionized by collisions with the target atoms (electron loss, denoted T, bottom part). Higher order collision sequences (P-T-P- ...) may occur ("Fermi shuttle", bottom part). The figure is taken from reference [1] with permission by the Royal Danish Academy of Science and Letters.

a cascade of "secondary electrons". Finally, some of the electrons liberated inside the solid reach the surface and may cross the surface potential barrier to be ejected into vacuum as summarized in figure 1a). Consequently, the emitted electron spectrum contains a rich variety of features including binary encounter electrons (BEE) and convoy electrons (CE), Auger electrons (AE), see figure 2, and the low-energy peak of "secondary" electrons (SE) in the eV range. BEE electrons are ejected with a velocity of $v_{BE} = 2\ v_p \cos(\theta)$ if the interaction with the target nucleus is neglected. For relativistic collisions, the momentum $p = \gamma v$ with $\gamma = (1-(v/c)^2)^{-1/2}$ is the relevant quantity instead of the projectile velocity v_p. The observed BEE peak reflects the initial momentum distribution of the bound target electrons ("Compton profile"). CE stem from either electron loss or capture to low-lying projectile continuum states. Their energy is close to zero in the projectile frame, and they form a characteristic cusp shaped peak in fast forward electron spectra. Furthermore, target AE (typically at low energies $E < 3$ keV), as well as projectile AE from in-flight de-excitation of heavy ions carrying electrons in bound excited states can be observed. Ejected electrons probe and provide information on interactions in different depths originating from different collision kinematics ranging from "soft" collisions and plasmon excitation to very "hard" close collisions. Most of the electrons have relatively low energy. We will -somewhat arbitrarily- consider "low energy electrons" as those that can typically be measured by charge collecting methods (for electron yields) or with a small electrostatic spectrometer (electron spectra, $0\ eV < E < 2$ keV).

The various experiments performed at GANIL aimed at measuring electron yields, and doubly differential yields (energy or velocity spectra at different ejection angles). They contributed decisively to our understanding of the very first step in energy deposition in matter, i.e. ionization and subsequent electron transport through condensed matter. One of the first experiments on electron emission from foils at GANIL showed that CE are very sensitive to the pre-equilibrium charge state evolution [2]. The emission of convoy (and BE-) electrons was also studied under channeling conditions [3]. More recently, new interest in electron emission under channeling conditions has arisen. The statistical distribution of the probability that n electrons are ejected P(n) (from which electron yields can be deduced) were measured in coincidence with the energy loss dE/dx and the emerging charge state of the projectiles [4]. Also, angular distributions of fast electrons were studied in connection with the detection of single heavy ions with CCD detectors [5] being used in radiobiological experiments.

2. Slow electrons

The first systematic studies of electron emission at CIRIL and GANIL were measurements of backward and forward electron yields as a function of carbon foil thickness [6-8]. These experiments allowed to establish the behaviour of electron emission as a function of the projectile velocity v_P, the projectile atomic number Z_P and target thickness d. Deduced simple scaling laws are important for applications, where often thin foils are used as electron source for e.g. beam monitors and dosimetry. These properties and scaling laws are discussed in [1]. Furthermore, the experimental data served as "benchmark" for numerical simulations of primary ionization and electron transport in solids [6, 7].

Low energy electron emission strongly depends on surface properties [9]. Therefore, experiments with sputter-cleaned surfaces in ultra-high vacuum are desirable and a specific set-up was built at CIRIL [10]. Electron energy spectra ($E < 1$ keV) and simultaneously measured electron yields were studied as a function of the projectile charge q_i at fixed velocity (9 MeV/u) [10-13]. Important results on "high charge effects" related to the strong perturbation in heavy ion collisions were obtained from those measurements and comparison to theoretical calculations of low energy electron spectra. The perturbation strength was varied from $q/v_P = 0.002$ (very weak perturbation, protons) to $q/v_P = 2$ (strong perturbation, GANIL heavy ions). The experimental results show a "reduction effect" with respect to a q^2 scaling for low energy electron emission under strong perturbation by heavy projectiles [10-12]. Such effects could arise from either specific high charge effects in primary ionization, or collective effects on electron transport in the vicinity of the projectile due to positively charged zone in the ion wake, which might attract the electrons moving away from the ion track [12, 14]. Theory rather points toward saturation of primary ionization cross sections with increasing ion charge [12, 15-16].

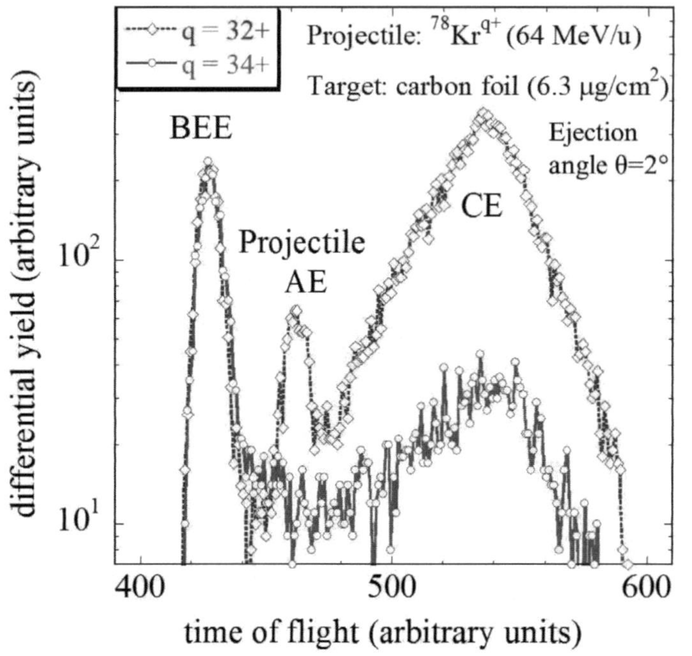

Figure 2. Fast forward electron spectra (the doubly differential electron yield) at small ejection angle θ=2° close to the beam direction as a function of the electron time of flight from impact of $^{78}Kr^{32+}$ and $^{78}Kr^{34+}$ (64 MeV/u) on a thin carbon foil. Prominent peaks are those of convoy electrons (CE, which with electron carrying projectiles mainly stem from projectile electron loss, denoted "T" in figure 1b) at longer time-of-flight (TOF) and of binary encounter electrons (BEE, denoted "P" in figure 1b) at shorter TOF. With $^{78}Kr^{32+}$ also Auger electrons (AE) from the in-flight de-excitation of the excited projectile are observed.

Another specific feature observed with heavy ions is the appearance of Auger hypersatellite lines from double ionization of the carbon K-Shell [11, 17]. From a careful analysis of the shape of carbon KVV Auger spectra, including background subtraction and deconvolution techniques, an increase of the width of the primary Auger electron spectrum with ion charge is observed. This effect is strongly connected to a modification of the electronic band structure of the solid target at the moment of the Auger decay. By adjusting the temperature-dependent Fermi distribution folded with the density of states to the observed primary line width, a mean "electronic temperature" in the track can be deduced [13, 17-19]. Measured "track temperatures" can thus test the predictions of thermal spike numerical simulations [17, 19] and serve as an in-site thermometer of the transient hot "plasma" inside the track. This experimental method gives quantitative information about the onset of the thermal spike. Information about the "track temperature" at later times and its evolution may be accessible via the measurement of the velocity of sputtered particles [20].

3. Fast electrons

Besides the above mentioned studies of CE, BE and δ-electrons [2, 3, 5] with the high energy GANIL beams, CE and BE were also studied at the medium energy facility SME [21, 22]. Also, by employing the multidetector array ARGOS, initially developed for studying nuclear reactions, and now adapted for detection of electrons (E > 10 keV) [23-27], ejection mechanisms of electrons at relativistic energies were studied in considerable detail. In comparison to conventional spectroscopic methods used in atomic collisions (i.e. electrostatic and magnetic analyzers), three important advantages were achieved: i) DDCS (velocity spectra) can be measured simultaneously at many ejection angles, ii) absolute cross sections can be measured with great accuracy thus allowing a stringent test of ionization theory iii) (multi-) coincidences can be measured. This latter feature opens the door for studies of processes which have not yet been accessible up to now (e.g. correlated emission of two or more electrons). By focusing on high energy BEE, a relativistic generalization of the electron impact approximation (EIA) and of free-electron transport could be successfully performed [22-25]. Measurements of electron spectra as a function of the target thickness allowed observing how electron emission evolves from single collisions (as in atomic collisions with low density gas targets) up to multiple collisions (as in the bulk of solids, where electron transport phenomena become important) and thus to link single collisions to effects in condensed matter as shown in figure 1a) [22-23, 25].

The study of high-energy electrons well beyond the binary encounter peak by the CIRIL group has received considerable international attention. A particularly interesting result is the observation of unexpectedly high cross sections for electron emission at energies far beyond the BE peak. Possibly, this can be explained by the so-called "Fermi-shuttle" process, which was introduced to explain the origin of high-energy cosmic radiation due to acceleration of charged particles by repeated collisions with moving magnetic fields in interstellar space. Multiple scattering sequences of electrons on the combined projectile-target systems can lead to electron acceleration (figure 1b) and emission of fast electrons even at large (backward) emission angles [28, 29].

Ongoing work concerns the application of electron spectroscopy in collisions of swift ions with insulating targets. The slowing down of BEE and CE allowed determining the target material dependent charging-up of polymer foils [30]. The angular dependence of in-flight emitted projectile Auger electrons with Kr^{32+} (64 MeV/u) [27] was also studied. For this collision system, the unique feature of the ARGOS multidetector, i.e. measurement of coincident emission of particles, allowed for the first time to study also the correlated emission of electrons. Under particular kinematic conditions (in-plane coincidence), a peak is observed which corresponds to an inelastic binary encounter with momentum transfer to an inner shell projectile electron. This is a rare ionization process with a specific kinematic signature involving a target and a projectile electron simultaneously, related to so-called (e,2e) experiments (one incoming electron, two outgoing electrons) [31]. Measurements of electron-X-ray coincidences are planned to study Radiative Electron Capture to the Continuum [32-33].

Acknowledgements
The experiments have been performed at GANIL. Financial support from the region of "Basse Normandie" is greatfully acknowledged. Numerous colleagues and co-authors from CIMAP-CIRIL, GANIL and other institutes contributed to this work. Thanks are due to all of them, and in particular to A. Billebaud, A. Clouvas, D. Dauvergne, K.O. Groeneveld, J.P. Grandin, S. Hagmann, D. Jakubassa-Amundsen, M. Jung, D. Lelièvre, V. Mouton, J.-M. Ramillon, S. Urso and C. Volant.

References
[1] Rothard H and Gervais B 2006 Electron emission from solids irradiated with swift ion beams *Ion Beam Science - Solved and Unsolved Problems* (*Mat. Fys. Medd. Dan. Vid. Selsk.* **52** 497-524) ed P Sigmund
[2] J. P. Gibbons J P, Elston S B, Kimura K, DeSerio R, Sellin I A, Burgdörfer J, Grandin J P, Cassimi A, Husson X, Liljeby L and Druetta M 1991 Observation of rapid evolution of convoy electron angular distributions *Phys. Rev. Lett.* **67** 481–484
[3] Andriamonje S, Chevallier M, Cohen C , Dural J , Genre R, Girard Y , Groeneveld K O, Kemmler J, Kirsch R, L'Hoir A, Maier R, Poizat J C, Quéré Y, Remillieux J, Schmaus D and Toulemonde M 1991 Charge exchange processes of high energy heavy ions channeled in crystals *Proceedings of the 4th Workshop on High Energy Ion-Atom-Collision Processes (Springer Lecture Notes in Physics vol 376)* ed D Berényi and G Kock pp 221-234
[4] Barrué F, Chevallier M, Dauvergne D, Kirsch R, Poizat J C, Ray C, Adoui L, Cassimi A, Rothard H, Toulemonde T, Cohen C, L'Hoir A, Vernhet D, Demonchy C, Giot L, Mittig W, Pita S, Roussel-Chomaz P and Billebaud A 2004 Electron emission induced by fast heavy ions in a thin silicon crystal *Phys. Rev.* A **70** 032902

[5] Meier MM, Schott J U, Strauch K and Groenveld K O 1998 Detection of δ-electron events in CCD: a fingerprint of single swift heavy ions *Nucl. Instrum. Meth. Phys. Res.* B **146** 601-606

[6] Rothard H, Caraby C, Cassimi A, Gervais B, Grandin J P, Jardin P, Jung M, Billebaud A, Chevallier M, Groeneveld K O and Maier R 1995 Target thickness dependent electron emission from carbon foils bombarded with swift highly charged heavy ions *Phys. Rev.* A **51** 3066-79

[7] Jung M, Rothard H, Gervais B, Grandin J P, Clouvas A and Wünsch R 1996 Transport of electrons induced by highly charged Ni (74 MeV/u) and Cu (9.6 MeV/u) ions in carbon: a study of target thickness dependent electron yields *Phys. Rev.* A **54** 4153-4161

[8] Rothard H, Jung M, Caron M, Grandin J P, Gervais B, Billebaud A, Clouvas A and Wünsch R 1998 Strong projectile dependent forward/backward asymmetry of electron ejection by swift heavy ions in solids *Phys. Rev.* A **57** 3660- 64

[9] Caron M, Rothard H and Clouvas A 2003 Influence of Adsorbates on Electron Emission from amorphous Carbon under Electron and swift heavy Ion Bombardment *Eur. Phys. J. Appl. Phys.* **21** 9-16

[10] Caron M, Rothard H, Jung M, Mouton V, Lelièvre D, Beuve M and Gervais B 1998 Mobile UHV Set-up for the Investigation of Electron Spectra at large Accelerators: first Results on High Charge Effects *Nucl. Instrum. Meth.* B **146** 126-130

[11] Caron M, Rothard H, Beuve M and Gervais B 1999 Multiple ionization and high charge effects in electron spectra from impact of 9 MeV/u ions on carbon *Phys. Scripta* T **80** 331

[12] Beuve M, Caron M, Gervais B and Rothard H 2000 Charge dependence of electron emission in swift heavy ion collisions with carbon *Phys. Rev.* B **62** 8818-8823

[13] Caron M, Rothard H, Beuve M and Gervais B 2001 Shape analysis of Auger electron spectra induced by highly charged ion impact on carbon *Phys. Scripta* T **92** 281-283

[14] Schiwietz G, Grande P, Skogvall B, Biersack J P, Köhrbrück R, Sommer K, Schmoldt A, Goppelt P, Kadar I, Ricz S and Stettner U 1992 Influence of nuclear track potentials in insulators on the emission of target Auger electrons *Phys. Rev. Lett.* **69**, 628-631

[15] Beuve M, Caron M, Fainstein P D, Galassi M, Gervais B, Rivarola R D and Rothard H 2002 Monte Carlo simulation of electron emission induced by swift highly charged Ions: beyond the linear response approximation *Eur. Phys. J.* D **21** 125-135

[16] Gervais B, Beuve M, Caron M and Rothard H 2003 Saturation effects in highly charged ion interaction with thin carbon foils *Nucl. Instrum. Meth. Phys. Res.* B **205** 835-840

[17] Schiwietz G, Xiao G, Luderer E and Grande P L 2000 Auger electrons from ion tracks *Nucl. Instrum. Meth.* B **164/165** 353-364

[18] Schiwietz G, G. Xiao G, Grande P L, Luderer E, Pazirandeh R, and Stettner U 1999 Determination of the electron temperature in the thermal spike of amorphous carbon *Europhys. Lett.* **47** 384-390

[19] Caron M, Rothard H, Toulemonde M, Beuve M and Gervais B 2006 Theoretical and experimental etudy of electronic temperatures in heavy ion tracks from Auger electron spectra and thermal spike calculations *Nucl. Instrum. Meth.* B **245** 36-46

[20] Hijazi H, Rothard H, Boduch P, Alzaher I , Langlinay T, Cassimi A , Ropars F, Been T, Ramillon J-M, Lebius H, Ban-d'Etat D, Farenzena L S and da Silveira E F 2012 Electronic sputtering of LiF by Krypton (10 MeV/u): size dependent energy distributions of $Li^+(LiF)_n$ clusters *Eur. Phys. J.* D **66**: 305

[21] Rothard H, Jakubassa-Amundsen D H and Billebaud A 1998 Solid State Effects in Binary Encounter Electron Emission from 13.6 MeV/u Ar^{17+} Collisions with C, Al, Cu and Au Foils *J. Phys.* B **31** 1563-1578

[22] Jakubassa-Amundsen D H and Rothard H 1999 Fast Electrons from Collisions of highly stripped Ions with solid-state Targets *Phys. Rev.* A **60** 385-391

[23] Rothard H, Lanzanò G, Jakubassa-Amundsen D H, De Filippo E, Mahboub D 2001 Theory and Measurement of absolute doubly differential Cross Sections of Binary Encounter Electron Ejection in Collisions of swift heavy Ions with Solids *J. Phys.* B **43** 3261-3274

[24] Lanzanò G, De Filippo E, Aiello S, Geraci M, Pagano A, Cavallaro S, LoPiano F, Pollaco E C, Volant C, Vuillier S, Beck C, Mahboub D, Nouicer R, Politi G, Rothard H and Jakubassa-Amundsen D H 1998 Fast Electron Production in Atomic Collisions induced by 77 A-MeV ^{40}Ar Ions studied with a Multidetector *Phys. Rev.* A **58** 3634-3641

[25] De Filippo E, Lanzanò G, Rothard H, Volant C, Jakubassa-Amundsen D H, Aiello S, Anzalone A, Arena N, Geraci M, Giustolisi F and Pagano A (2003): Absolute Cross Sections for Binary-Encounter Electron Ejection by 95 MeV/u $^{36}Ar^{18+}$ penetrating Carbon Foils *Phys. Rev.* A **68** 024701

[26] De Filippo E, Lanzanò G, Rothard H, Volant C, Aiello S, Anzalone A, Arena N, Geraci M, Giustolisi F, and Pagano A 2004 Fast Electron Ejection from C, Ni, Ag and Au Foils by $^{36}Ar^{18+}$ (95 MeV/u): Measurement of absolute Cross Sections *Eur. Phys. J.* A **21**169-174

[27] Lanzanò G, De Filippo E, Hagmann S, Rothard H and Volant C 2007 In flight emission of projectile Auger electrons from highly energetic heavy ions (20 MeV/u$<$E$<$100 MeV/u) colliding with carbon foils *Nucl. Instrum. Meth.* B **256** 510-514

[28] Sulik B, Stolterfoht N, Hellhammer R, Pésiç Z, Koncz C, Tőkési K and Berényi D 2003 Fermi-shuttle acceleration of electrons in ion–matter interaction *Nucl. Instrum Meth.* B **212** 33-44

[29] Rothard H, Lanzanò G, De Filippo E and Volant C 2005 Fermi Shuttle Acceleration in atomic Collisions: the Case of ion induced Electron Emission *Nucl. Instrum. Meth.* B **230** 419-425

[30] De Filippo E, Lanzano G, Amorini F, Geraci E, Grassi L, La Guidara E, Lombardo I, Politi G, Rizzo F, Russotto P, Volant C, Hagmann S and Rothard H 2011 Slowing down of fast electrons as probe for charging and decharging dynamics of ion-irradiated insulators *Phys. Rev.* A **83** 064901

[31] De Filippo E, Lanzanò G and Rothard H 2007 Impact and Applications: Fast Electron Ejection in Atomic Collisions Studied with the ARGOS Multi-Detector *Nucl. Phys. News.* **17** No. 2 24-27

[32] Hillenbrand P M, Hagmann S, Stoehlker T, Litvinov Y, Kozhuharov C, Spillmann U, Shabev V M, Stiebing K, Lestinsky M, Surzhykov A, Voitkiv A, Franzke B, Fischer D, C. Brandau C, Schippers C S, Mueller A, Schneider D, Jakubassa D, Artiomov A, De Filippo E, Ma X, Dörner R and Rothard H 2013 Future experiments using forward electron spectroscopy to study quantum dynamics of high-Z ions at the ESR/CRYRING storage rings *Phys. Scripta.* T **156** 014087

[33] Hillenbrand P M, Hagmann S, Atanasov D, Banaś D, Blumenhagen K-H , Brandau C, Chen W, De Filippo E, Gumberidze A, Guo D L, Jakubassa-Amundsen D H, Kovtun O, Kozhuharov C, Lestinsky M, Litvinov Yu A, Müller A, Müller R A, Rothard H, Schippers S, Schöffler M S, Spillmann U, Surzhykov A, Trotsenko S, Winckler N, Yan X L, Yerokhin V A, Zhu X L and Stöhlker T 2014 Radiative-electron-capture-to-continuum cusp in U^{88+}+N_2 collisions and the high-energy endpoint of electron-nucleus bremsstrahlung *Phys. Rev.* A **90** 022707

Radiation effects in astrophysical ices

Philippe Boduch[1], Emmanuel Dartois[2,3], Ana L. F. de Barros[4], Enio F. da Silveira[5], Alicja Domaracka[1], Xue-Yang Lv[1], Maria Elisabetta Palumbo[6], Sergio Pilling[7], Hermann Rothard[1], Eduardo Seperuelo Duarte[8] and Giovanni Strazzulla[6]

[1] Centre de Recherche sur les Ions, les Matériaux et la Photonique (UMR6252), CEA/CNRS/ENSICAEN/Université de Caen-Basse Normandie UCBN, CIMAP-CIRIL-Ganil, BP 5133, 14070 Caen Cedex 05, France
[2] CNRS-INSU, Institut d'Astrophysique Spatiale, UMR 8617, 91405 Orsay, France
[3] Université Paris Sud, Institut d'Astrophysique Spatiale, UMR 8617, bâtiment 121, 91405 Orsay, France
[4] Departamento de Física, Centro Federal de Educacão Tecnológica Celso Suckow da Fonseca, Av. Maracanã 229, 20271-110 Rio de Janeiro, RJ, Brazil
[5] Departamento de Física, Pontifícia Universidade Católica do Rio de Janeiro, Rua Marquês de São Vicente 225, 22451-900 Rio de Janeiro, RJ, Brazil
[6] INAF–Osservatorio Astrofisico di Catania, Catania, Italy
[7] Universidade do Vale do Paraiba (UNIVAP), Instituto de Pesquisa e Desenvolvimento (IP&D), Av. Shishima Hifumi 2911, CEP 12244-000 São José dos Campos, SP, Brazil
[8] Grupo de Física e Astronomia, Instituto Federal de Educação Ciência e Tecnologia do Rio de Janeiro, Rua Lucio Tavares, 1045, 26530-060, Nilópolis, RJ, Brazil

E-mail: rothard@ganil.fr

Abstract. The interaction of heavy ions with astrophysical ices was studied at different beamlines of GANIL by infrared absorption spectroscopy. This allowed simulating in the laboratory the physico-chemical modifications induced in icy objects in space, exposed to radiation fields such as the solar wind, magnetospheric particles and interstellar cosmic rays. We briefly discuss sputtering, destruction and formation of molecules, amorphization and compaction, implantation, and finally the formation of organic molecules. This latter topic is related to the question of the initial conditions for the emergence of life.

1. Introduction

Ices are ubiquitous in space: in the dense phases of the interstellar medium (dense molecular clouds and proto-stellar cores, shown in figure 1) and in the solar system. The molecules can be formed or freeze onto surfaces of dust grains and form thin icy mantles. In the Oort cloud and the Kuiper belt of the solar system, together with dust (carbon based materials, silicates), they can form icy bodies, some of them being the progenitors of comets. They can cover the surfaces of the giant planets' satellites. Simple molecules such as H_2O, CO, CO_2, NH_3, CH_4 etc. are the dominant species, but more complex, organic molecules (alcohols etc.) have been detected. The ices are continuously exposed to ionizing irradiation (UV photons, electrons, ions); this may induce several physico-chemical processes. Among

Journal of Physics: Conference Series **629** (2015) 012008 doi:10.1088/1742-6596/629/1/012008

them are fragmentation/radiolysis, followed by chemical reactions and formation of new molecules; structural/phase changes; and desorption/sputtering of molecules from the surface [1]. In astrophysical studies on interaction of radiation with ices, effects induced by weakly ionizing radiation such as UV photons and keV-MeV light ions (H, He) was extensively studied in the laboratory. This was driven by the use of keV to MeV proton accelerators, that are relatively small and more common than heavy ion accelerators such as GANIL, which are indeed large scale installations of which only a few exist in the world. Here, we therefore focus on the contribution of heavier ions (C, O, S, Fe), which are also present in space.

Figure 1. Herschel's infrared view of part of the Taurus Molecular Cloud, which is about 450 light-years from Earth and is the nearest large region of star formation. Image credits: ESA/Herschel/SPIRE (reproduced with permission).

From a collaboration between PUC-Rio and CIMAP-CIRIL with the goal to study sputtering of LiF by MeV ion beams, a subject under intense research in both laboratories (see [3] and references therein), emerged a new research topic at GANIL: heavy ion induced radiation effects in astrophysical ices. Indeed, at PUC, the investigation of sputtering of secondary ions from ices by MeV heavy ions started a decade ago [4, 5]. It was observed that the dependence of the ion desorption yield on the projectile stopping power was strong enough that it can compensate for the low abundance of heavy ion constituents of cosmic rays and, therefore, produce effects in solids comparable to those induced by protons and alpha particles. The sputtering occurs basically through neutrals, whose yields are orders of magnitude higher than that of secondary ions.

It was then realized that a new method could be coupled for sputtering yield measurement of ices: Fourier Transform Infrared Spectroscopy (FTIR) was able not only to identify formed molecules in the target and measure their formation and destruction cross sections, but also to measure total sputtering yields. Indeed, following the decrease of peak areas of infrared bands during ion beam bombardment and compensating for the newly created species, the removal rate of material could be determined. Consequently, in close collaboration between CIMAP/Caen, IAS/Orsay and PUC/Rio de Janeiro, the radiolysis and sputtering of ices such as CO and CO_2 by swift heavy ions (cosmic ray analogs) at about 15K was studied. The research was strongly supported by the French-Brazilian CAPES-COFECUB exchange program. Since 2010, INAF/Catania joined, and, supported by the COST Action "The Chemical Cosmos", the studies were extended to effects induced by slow heavy ions (solar wind and giant planet's magnetospheres).

Journal of Physics: Conference Series **629** (2015) 012008 doi:10.1088/1742-6596/629/1/012008

2. Cosmic rays

The first laboratory simulations of the interaction of heavy cosmic ray ions with ices were performed within a joint-doctorate program between the PUC-Rio de Janeiro and the University of "Caen-Basse Normandie". E. Seperuelo Duarte, graduate from Astronomy, was able to apply the new approach in the study of CO_2 ice sputtering by 46 MeV Ni at the IRRSUD beam line of GANIL [6]. The experiment was repeated for CO ice sputtering induced by 50 MeV Ni (again at IRRSUD) and by 537 MeV Ni (at the SME beamline) [7]. A strong non-linear increase of sputtering yields Y with deposited energy in CO ice due to electronic sputtering was observed as shown in figure 2. The quadratic dependence with the electronic stopping power, $Y \sim S_e^2$, is in excellent agreement with previous results obtained with light ions, at lower energies, by Brown et al. [8]. The stopping power $S_e = -dE/dx$ of a material is equal to the energy E deposited per unit path length x by the projectile.

Therefore, even taking into account the low abundance, the total number of heavy ion desorbed/sputtered molecules is higher than that desorbed by protons. This observation now needs to be taken into account for understanding the evolution of astronomical objects such as dense molecular clouds (figure 1). The heavy ion induced desorption contribute to explain for the observed presence of gas phase molecules in such clouds. In the absence of these processes, it is expected that all CO molecules should condense on small grains in cold (10 K) dense clouds within timescales shorter than these clouds ages, and form an icy mantle. Processes discussed to explain the gas phase molecules include photodesorption of ices by secondary UV photons from cosmic ray interaction with hydrogen or "spot heating" of grains, not to be confused with our observation related to electronic sputtering [7]. Deep inside dense clouds, each of the three processes (electronic desorption by the heavy ion fraction of cosmic rays, and "spot heating" of grains by cosmic rays, photodesorption) yield comparable contributions to the total desorption yields [7].

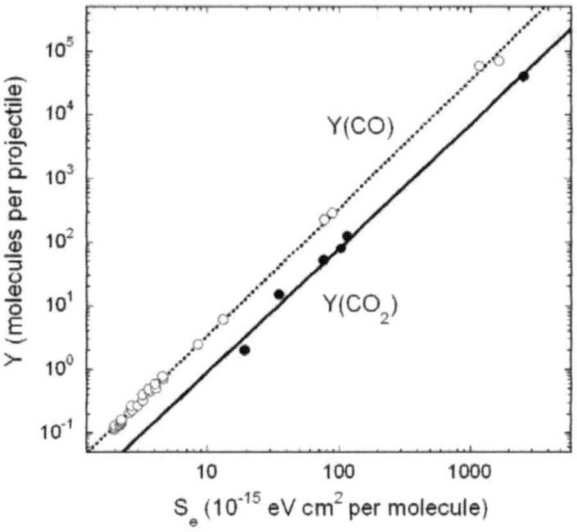

Figure 2. Sputtering yield Y of CO and CO_2 as a function of electronic stopping power S_e. A quadratic dependence $Y \sim S_e^2$ is observed [6-9].

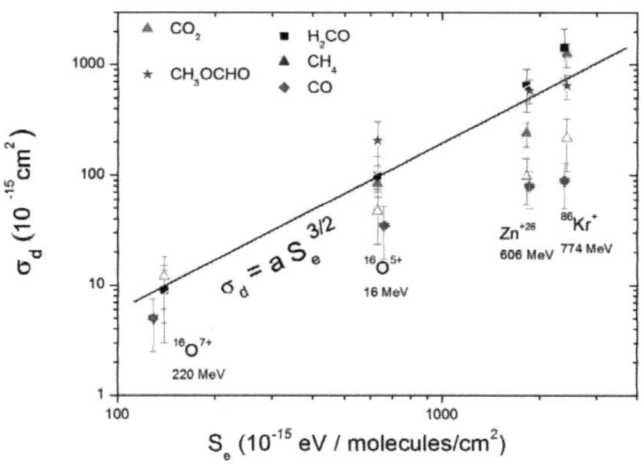

Figure 3. Destruction cross sections σ_d of ice molecules (as indicated) as a function of electronic stopping power (see text); they essentially follow a power law $\sigma_d \sim S_e^n$ with $n \approx 1.5 \pm 0.3$ [14, 17, 21].

Furthermore, heavy ions are an order of magnitude more efficient in destroying the initial molecules and corresponding radiochemical yields than protons. A quadratic dependence of the sputtering yield on stopping power is not only observed for CO ices [7, 8], but also for CO_2 ices [7, 9] (figure 2) and for high electronic stopping also for O_2, N_2 and H_2O [9] thus supporting the conclusions concerning the importance of heavy ions in modification of interstellar ices.

Journal of Physics: Conference Series **629** (2015) 012008 doi:10.1088/1742-6596/629/1/012008

Since then, cosmic ray effects in many other ices have been analyzed at CIMAP-GANIL: H_2O [10-12], NH_3 [13, 14], CH_4 [15, 16], CH_3OH [17, 18], HCOOH [19-20], H_2CO [21], C_6H_{12} [22] and acetone [23]. Chemical reactions follow the fragmentation of initial molecules. This leads to formation of new species, a finding particularly interesting with ice mixtures containing the ingredients of organic molecules (H,C,N,O) [13, 22, 24, 25]. The determination of destruction and formation cross sections is one of the most important goals of this research. Considering the obtained results with low energy data, we proposed that – at least for ice targets – such cross sections depend basically only on a power ($\approx 1.5 \pm 0.3$) of the electronic stopping [14, 17, 18, 21]. As a consequence, it is possible to estimate molecular destruction rates for all cosmic ray constituents over a very large energy range. Figure 3 illustrates this "law" for methanol ice. Moreover, it turns out from these calculations that - in interstellar medium – chemical evolution induced by iron ions of cosmic rays is surprisingly faster than that induced by protons and alpha particles: for instance, the methanol half live due to iron bombardment is 3×10^{14} s (approx. 10^7 years), 700 times shorter than that induced by protons.

Often neglected compared to radiolysis induced chemical modification, the physical state of the ice (amorphous, crystalline, metastable) is also extremely important in many aspects for astrophysicists. At low temperature (about 10 K), a porous amorphous ice phase is obtained by accretion of water molecules, which is getting compacted under ion irradiation. Finally, it evolves towards a compact non-porous, but disordered amorphous phase [26]. Irradiation of crystalline water ice annealed at "high" temperature (140 K) and cooled down to approximately 10 K leads to amorphization towards this same intermediate amorphous phase [27].

The compaction, i.e. the loss of open porosity of amorphous ice films was studied at the high energy beam line LISE of GANIL as well as at SME and IRRSUD in the MeV to GeV range [10]. FTIR allows to study the number of "OH dangling bonds" (i.e. pending water molecule OH bonds not engaged in a hydrogen bond in the initially porous ice structure). Combined with previous experiments using lower energy light ions from other facilities, we evaluated a cross section for the disappearance of porosity as a function of electronic stopping power over three orders of magnitude. The relevant phase structuring time scale for the ice network on icy mantles covering interstellar dust grains are present in the molecular clouds of our galaxy was compared to interstellar chemical time scales using an astrophysical model including the expected galactic cosmic ray distribution of abundances. The galactic cosmic ray distribution of ion abundances reproduces the ionization rate observed and provides ice compaction time scales in the range from 1.4×10^5 to 2×10^6 years. Ice mantle porosity or pending bonds as monitored by the OH dangling bonds, are removed efficiently by cosmic rays. Studies on the compaction and amorphization of water ice are ongoing. Similar observations have also been made for e.g. H_2O-NH_3 and other ices made of mixtures of molecules [13, 14].

3. Ion Implantation

When the penetration depth of an energetic ion that travels through a material is shorter than the thickness of the target, the ion finally is stopped and remains implanted. Implantation experiments are particularly relevant to understand the effects of different ion populations such as solar wind ions and magnetospheric particles, interacting with icy objects in the Solar System (for a review see [28]). Experiments with reactive ions (e.g., H^+, C^+, N^+, O^+, S^+) are particularly relevant because they induce all of the effects of any other ion, but in addition have a chance, by implantation in the target, to form new species containing the projectile. In recent years several implantation experiments have been conducted at the low energy facility ARIBE of GANIL. The experiments were performed with carbon and sulfur ions of energies between 10 keV and 180 keV. They are able to break 10^3-10^4 bonds per ion and thus a large number of new molecules can be formed by recombination of fragments of the irradiated species. In addition, implanted carbon and sulfur, being reactive, have a chance to form species that include the projectile with a maximum yield of one molecule per incoming ion. Recently, several experiments ([29-31], listed in table 1) have been conducted at ARIBE.

Journal of Physics: Conference Series **629** (2015) 012008 doi:10.1088/1742-6596/629/1/012008

Table 1. Summary of the implantation experiments conducted so far at ARIBE.

ION (E in keV)	TARGET T= 15-150 K	MAJOR PRODUCED SPECIES (in bold those containing the projectile)	REF
C^{q+} (10, 30) q=2-3	H_2O	H_2O_2, $\mathbf{CO_2}$	[29]
S^{q+} (35-180) q= 7, 9, 11	H_2O	H_2O_2 $\mathbf{H_2SO_4}$ dissolved in H_2O	[30]
	CO	$\mathbf{SO_2}$	[31]
	CO_2	$\mathbf{SO_2}$	[31]

Figure 4. Artist's impression of JUICE, the JUpiter ICy moons Explorer mission, in the Jovian system. Image credits: ESA/AOES (reproduced with permission).

The results have mostly been applied to the icy moons of the giant planets in the outer Solar System, where in fact water ice is by far the dominant species. Prominent examples are the Jovian moons Europa, Callisto and Ganymede, which have been visited by the Galileo spacecraft and which are the focus of Europe's ESA next large science mission, the Jupiter Icy moons Explorer JUICE (figure 4). On these objects, several absorption features have been observed and have been attributed to C-H, H_2O_2, S-H (and/or H_2CO_3), SO_2, CO_2, and CN [32, 33].

Journal of Physics: Conference Series **629** (2015) 012008 doi:10.1088/1742-6596/629/1/012008

In addition abundant surface constituents of the Galilean moons are darker, non-ice materials [34], and in the case of Europa, hydrated sulfuric acid [35]. The dark materials are thought to be carbonaceous compounds present at the time of the satellite formation or later delivered by cometary and meteoritic bombardment [36]. Results based on space observations have shown that most of the CO_2 detected on the surfaces of Callisto and Ganymede is contained in the non-ice materials [33, 34]. SO_2 is believed to be present on the surface of the icy moons as a molecule trapped in the dark material and/or the water ice [32]. Many of the icy moons orbit within their planet's magnetosphere and their surfaces are subjected to intense bombardment. The major components of the Jovian magnetosphere are protons and ions such as S^{q+} and O^{q+}, and also energetic electrons [39].

The high fluxes of those ions justify the need of implantation experiments that are conducted to contribute to the debate on the origin of the observed minor molecular species: are they native of the given satellite (endogenic source) or are they due to external effects such as ion implantation (exogenic source)? The experiments conducted at ARIBE [29-31] have given an important contribution to the discussion. As an example, in figure 5 we show the IR spectra of water ice before and after implantation of 3×10^{15} 30 keV $^{13}C^{3+}/cm^2$ (left panel) and 4.8×10^{16} 105 keV S^{7+}/cm^2 (right panel). A band appearing at about 2275 cm^{-1} upon implantation can easily be attributed to $^{13}CO_2$ (left panel of figure 5). The corresponding measured yields for ions with energies from 10 keV to 30 keV do neither depend on the charge state (q = 1-3) nor on the temperature (15 K and 80 K). They range between 0.32 and 0.57 molecules/ion, i.e. they are constant within the experimental error. The time scale necessary to accumulate the observed quantity of carbon dioxide on the surface of Europa by implantation of magnetospheric carbon ions has been evaluated to be on the order of $(1.0\text{-}1.3) \times 10^4$ years. This time is higher than that evaluated for the production of carbon dioxide by other relevant processes, in particular by the ion bombardment stimulated formation of CO_2 at the interface between solid carbon and water ice [40]. We therefore conclude that although a relevant quantity of CO_2 can be formed by carbon ion implantation, this is not the dominant formation mechanism on the surfaces of icy satellites [29].

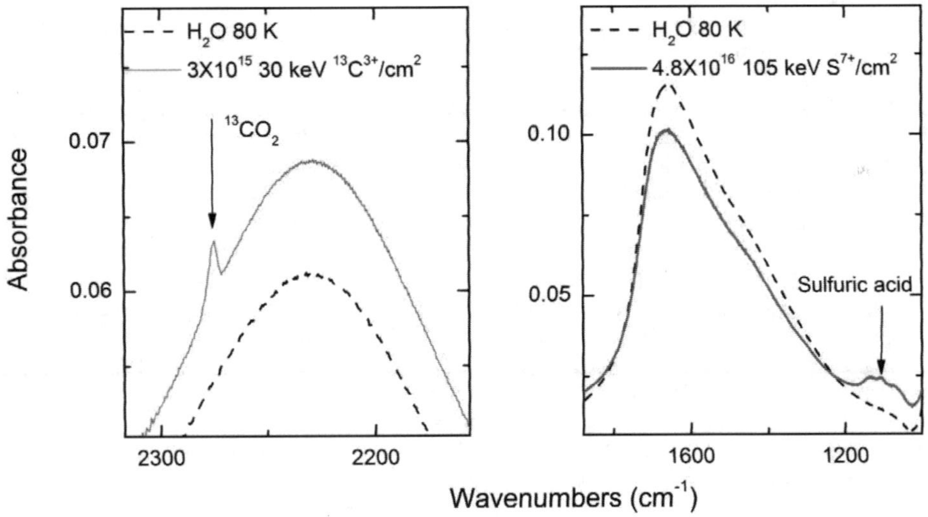

Figure 5. IR spectra of water ice before and after implantation of carbon ions (left panel) and sulfur (right panel) ions.

The right panel of figure 5 shows that a new, multi-peaked band is formed around 1150 cm^{-1}, indicating the formation of hydrated sulfuric acid upon sulfur implantation. We have performed experiments with sulfur ions having different charges (q = 7, 9, 11) and energies (35 keV - 176 keV), implanted in water ice at 80 K. The results of our experiments indicate that the formation yields increase with ion energy, from 0.12 molecules/ion for 35 keV ions to 0.64 molecules/ion for 200 keV ions (this latter value is from [41]). From the high sulfur ion fluxes of $(2\times10^{6}\text{-}1\text{x}10^{8})$ ions cm^{-2} s^{-1}) evaluated in different regions of the surface of Europa [42], Ding et al. [30] conclude that sulfur ion implantation is the dominant formation mechanism of hydrated sulfuric acid. The suggestion that the observed distribution of sulfuric acid on the surface is well correlated with the local flux of sulfur ions [42] finds a full explanation by our experimental data.

Nevertheless, the question about the origin of SO_2 remains open. We were not able to find any evidence of synthesis upon sulfur ion implantation into water ice. In fact, SO_2 has been observed on some of the icy moons by ultraviolet observations, via its band peaked at 280 nm. To explain its origin several mechanisms have been proposed, among which S implantation in water ice [43, 44], which is not supported by our experimental results. Further experimental efforts are thus necessary to clarify the question.

4. Outlook: formation of organic pre-biotic molecules

The matter in our galaxy evolves within a life cycle of stars, which depends on numerous astrophysical processes including processing by radiation fields and astrochemical processes: radiation effects, reactions at gas-grain surfaces, ice photochemistry and so on. Interstellar dust grains, after the collapse of dense molecular clouds are incorporated in protoplanetary systems. Thus, interstellar matter can contribute to the so-called primitive matter. Later on, comets (orbiting in the Oort cloud, the Kuiper belt or Jupiter family ones, made out of ices and dust) and asteroids, can feed this matter to planets such as the early Earth. Laboratory astrochemistry has shown that complex molecules (including amino acids and nucleobases) can be formed under interstellar/circumstellar conditions by vacuum UV irradiation, electron and ion bombardment of interstellar ice analogs containing small molecules (H_2O, CO, CO_2, CH_3OH, NH_3, HCOOH), that is, the basic building blocks of organic matter (H, C, N, O, and others). Therefore, irradiation processing of ice mixtures [13, 22, 24, 25, 45] may be related to the question of the origin of organic matter on Earth, bringing potential building blocks for the emergence of life on earth, and possibly on other bodies in the solar system.

At IRRSUD, ammonia-containing ices made of mixtures of small molecules such as H_2O:NH_3 and H_2O:NH_3:CO were irradiated by 46 MeV Ni ions in an attempt to simulate the physical chemistry induced by heavy-ion cosmic rays inside dense astrophysical environments. The infrared spectra of the irradiated ice samples exhibit lines of several new species including HNCO, N_2O, OCN^-, and NH_4^+. After FTIR measurements as a function of the projectile fluence in order to determine destruction- and formation cross sections, the irradiated samples were slowly warmed up to room temperature. In the case of the irradiated H_2O:NH_3:CO ice, the infrared spectrum at room temperature contains five bands that can be tentatively assigned to vibration modes of zwitterionic glycine, an amino acid, and possibly, another can be attributed to hexamethylenetetramine (HMT) [13]. If such organic molecules exist in space, it is also pertinent to study their stability when exposed to radiation [46].

Furthermore, ices containing methanol and ammonia CH_3OH:NH_3 were irradiated at IRRSUD. The main goal here was to make the first proper comparison between the formation of products in an ice mixture of interstellar type both by irradiation of the ice using swift heavy ions and by vacuum UV photons using a similar energy dose (in eV/molecule). It was shown that both UV photon and swift heavy ion irradiation yield similar residues formed after warm-up to room temperature of the irradiated ices [45]. In order to confirm and quantify the possible formation of even more complex organic molecules, in the near future, new irradiation experiments with ex-situ chromatographic analysis of thick residues from swift ion processing are in preparation. This could lead to the detection of new molecular components in the organic residues. These species could thus be present in circumstellar regions and comets, and were likely delivered to the primitive Earth.

Studies on radiation effects in astrophysical ices will benefit in the near future from a new ultrahigh vacuum setup IGLIAS financed by ANR (partners: CIMAP and IAS) allowing to work under improved experimental conditions. In view of the increasing amount of data arising from ever improved observation tools including earth and satellite based telescopes, and space probes such as JUICE, dedicated laboratory experiments are needed to interpret the observations. Well established and emerging experimental techniques [47] will help to understand formation of dust grains, icy mantles and their processing dust grain formation and processing (ice mantle chemistry, space weathering [48]) under astronomical conditions.

Acknowledgements

The experiments have been performed at GANIL. This work has been supported by PHC Capes-Cofecub France-Brésil, EU Cost Action "The Chemical Cosmos", Chinese Scholarship Council CSC, Région Basse Normandie, SPIRIT (EU) and EMIR (France) networks, the French INSU-CNRS program Physique et Chimie du Milieu Interstellaire (PCMI) and the ANR IGLIAS, grant ANR-13-BS05-0004 of the French Agence Nationale de la Recherche. It is a pleasure to thank our colleagues from CIMAP and GANIL for their invaluable support. Special thanks to E. Balanzat, T. Been, A. Cassimi, F. Durantel, S. Guillous, C. Grygiel, P. Guinement, O. Kamalou, D. Lelièvre, F. Levesque, M. Lion, T. Madi, L. Maunoury, I. Monnet, Y. Ngono-Ravache, F. Noury, J. M. Ramillon, F. Ropars, J. C. Thomas, and the LISE team. Special thanks also to all co-authors and in particular D. P. P. Andrade, D. Baklouti, V. Bordalo, M. Chabot, J. J. Ding, L. S. Farenzena, D. Fulvio, T. Langlinay, C. F. Mejía Guamán, G. M. Muñoz Caro and R. Brunetto.

References

[1] Allamandola L J, Bernstein M P, Sandford S A and Walker R L 1999 Evolution of interstellar Ices *Space Sci. Rev.* **90** 219–232

[2] Shen J C, Greenberg J M, Schutte W A and Van Dishoek E F 2004 Explosive chemical desorption in dense clouds *Astron. Astrophys.* **415** 203-215

[3] Hijazi H, Farenzena L S, Rothard H, Boduch P, Grande P L and da Silveira E F 2011 Cluster ion emission from LiF induced by MeV N^{q+} projectiles and ^{252}Cf fission fragments *Eur. Phys. J.* D **63** 391-400

[4] Collado V M, Farenzena L S, Ponciano C R, da Silveira E F and Wien K 2004 Ion desorption from frozen H_2O irradiated by MeV heavy ions *Surf. Sci.* **569** 149-162

[5] Farenzena L S, Iza P, Martinez R, Fernandez-Lima F A, Seperuelo Duarte E, Faraudo G, Ponciano C R, Homem M P G, Naves de Brito A, Wien K and da Silveira E F 2005 Electronic sputtering analysis of astrophysical ices *Earth Moon Planets* **97** 311-329

[6] Seperuelo Duarte E, Boduch P, Rothard H, Been T, Dartois E, Farenzena L S and da Silveira E F 2009 Heavy ion irradiation of condensed CO_2: sputtering and molecule formation *Astron. Astrophys.* **502** 599-603

[7] Seperuelo Duarte E, Domaracka A, Boduch P, Rothard H, Dartois E, and da Silveira E F 2010 Laboratory simulation of heavy ion cosmic ray interaction with condensed CO *Astron. Astrophys.* **512** A71

[8] Brown W L, Augustyniak W M, Marcantonio K J, Simmons E H, Boring J W, Johnson R E and Reimann C T 1984 Electronic sputtering of low temperature molecular solids *Nucl. Instrum. Meth.* B **1** 307-314

[9] Raut U and Baragiola R A 2013 Sputtering and molecular synthesis induced by 100 keV protons in condensed CO_2 and relevance to the outer Solar System *Astrophys. J.* **772** 53

[10] Dartois E, Ding J J, de Barros A L F, Boduch P, Brunetto R, Chabot M, Domaracka A, Godard M, Lv X Y, Mejía Guamán C F, Pino T, Rothard H, da Silveira E F and Thomas J C 2013 Swift heavy ions irradiation of water ice at MeV to GeV energies - Approaching true cosmic ray compaction *Astron. Astrophys.* **547** A97

[11] Boduch P, da Silveira E F, Domaracka A, Gomis O, Lv X Y, Palumbo M E, Pilling S, Rothard

H, Duarte E S and Strazzulla G 2011 Production of oxidants by ion bombardment of icy moons in the outer Solar System *Adv. Astron.* Article ID 327641

[12] Pilling S, Seperuelo Duarte E, Domaracka A, Rothard H, Boduch P and da Silveira E F 2010 Radiolysis of H_2O:CO_2 ices by energetic and heavy solar wind particles and cosmic ray analogs *Astron. Astrophys.* **523** A77

[13] Pilling S, Seperuelo Duarte E, da Silveira E F, Balanzat E, Rothard H, Domaracka A and Boduch P 2010 Radiolysis of ammonia-containing ices by energetic, heavy and highly charged ions inside dense astrophysical environments *Astron. Astrophys.* **509** A87

[14] Bordalo V, da Silveira E F, Lv X Y, Domaracka A, Rothard H, Seperuelo Duarte E and Boduch P 2013 Chemical processing of pure ammonia and ammonia-water ices induced by heavy ions *Astrophys. J.* **774** 105

[15] de Barros A L F, Bordalo V, Seperuelo Duarte E, da Silveira E F, Domaracka A, Rothard H and Boduch 2011 Cosmic ray impact on astrophysical ices: Laboratory studies on heavy ion irradiation of methane *Astron. Astrophys.* **531** A160

[16] Mejía C F, de Barros A L F, Bordalo V, da Silveira E F, Boduch P, Domaracka A and Rothard H 2013 Radiolysis of methane ice at 15 K by MeV O, Fe and Zn ions *Mon. Not. R. Astron. Soc.* **433** 2368–2379

[17] de Barros A L F, Domaracka A, Andrade D P P, Boduch P, Rothard H and da Silveira E F 2011 Radiolysis of frozen methanol by heavy cosmic ray and energetic solar particle analogues ions *Mon. Not. R. Astron. Soc.* **418** 1363–1374

[18] de Barros A L F, da Silveira E F, Rothard H, Langlinay T and Boduch P 2014 Chemical and physical effects induced by heavy cosmic ray analogs on frozen methanol and water ice mixtures ions *Mon. Not. R. Astron. Soc.* **443** 2733-2745

[19] Andrade D P P, de Barros A L F, Pilling S, Domaracka A, Rothard H, Boduch P and da Silveira E F 2013 Chemical reactions induced in frozen formic acid by heavy ion cosmic rays ions *Mon. Not. R. Astron. Soc.* **430** 787-796

[20] Bergantini A, Pilling S, Rothard H, Boduch P and Andrade D P P 2014 Processing of formic acid-containing ice by heavy and energetic cosmic ray analogues ions *Mon. Not. R. Astron. Soc.* **437** 2720–2727

[21] de Barros A L F, da Silveira E F, Pilling S, Domaracka A, Boduch P and Rothard H 2014 Processing of low content carbon interstellar ice analog by cosmic rays: Implications on the chemistry around oxygen-rich stars ions *Mon. Not. R. Astron. Soc.* **438** 2026-2035

[22] Pilling S, Andrade D P P, da Silveira E F, Rothard H, Domaracka A and Boduch P 2012 Formation of unsaturated hydrocarbons in interstellar ice analogs by cosmic ray ions *Mon. Not. R. Astron. Soc.* **423** 2209–2221

[23] Andrade D P P, A. L. F. de Barros A L F, Ding J J, Rothard H, Boduch P and da Silveira E F 2014 Acetone degradation by cosmic rays in the solar neighborhood and in the galactic centre ions *Mon. Not. R. Astron. Soc.* **444** 3792 -3801

[24] Bernstein M P, Dworkin J P, Sandford S A, Cooper G W and Allamandola L J 2002 Racemic amino acids from the ultraviolet photolysis of interstellar ice analogues *Nature* **416** 401-403

[25] Muñoz Caro G M, Meierhenrich U J, Schutte W A, Barbier B, Arcones Segovia A, Rosenbauer H, Thiemann W H-P, Brack A and Greenberg J M 2002 Amino acids from ultraviolet irradiation of interstellar ice analogues *Nature* **416** 403-406

[26] Palumbo M E 2006 Formation of compact solid water after ion irradiation at 15 K *Astron. Astrophys.* **453** 903-909

[27] Fama M, Loeffler M J, Raut U and Baragiola R A 2010 Radiation-induced amorphization of crystalline ice *Icarus* **207** 314-319

[28] Strazzulla G 2011 Cosmic ion bombardment of the icy moons of Jupiter *Nucl. Instrum. Meth.* B **269** 842–851

[29] Lv X Y, de Barros A, Boduch P, Bordalo V, da Silveira EF, Domaracka A, Fulvio D, Hunniford CA, Langlinay T, Mason N, McCullogh R, Palumbo M E, Pilling S, Rothard H and

Strazzulla G 2012 Implantation of multiply charged carbon ions in water ice *Astron. Astrophys.* **546** A81

[30] Ding J J, Boduch, P Domaracka A, Guillous S, Langlinay T, Lv X Y, Palumbo M E, Rothard H and Strazzulla G 2013 Implantation of multiply charged sulfur ions in water ice *Icarus* **226** 860

[31] Lv X Y, Boduch P, Ding J J, Domaracka A, Langlinay T, Palumbo M E, Rothard H and Strazzulla G 2014 Sulfur implantation in CO and CO_2 ices ions *Mon. Not. R. Astron. Soc.* **438** 922-929

[32] McCord T B, Carlson R W, Smythe W D, Hansen G B, Clark R N, Hibbitts C A, Fanale F P, Granahan J C, Segura M, Matson D L, Johnson T V and Martin P D 1997 Organics and other molecules in the surfaces of Callisto and Ganymede *Science* **278** 271

[33] Hage W, Liedl K R, Hallbrucker A and Mayer E 1998 Carbonic acid in the gas phase and its astrophysical relevance *Science* **279** 1332

[34] McCord T B, Hansen G B, Clark R N, Martin P D, Hibbitts C A, Fanale F P, Granahan J C, Segura M, Matson D L, Johnson T V, Carlson R W, Smythe W D and Danielson G E 1998 Non-water-ice constituents in the surface material of the icy Galilean satellites from the Galileo near-infrared mapping spectrometer investigation *J. Geophys. Res.* **103** 8603

[35] Carlson R W, Johnson R E and Anderson M S 1999 Sulfuric acid on Europa and the radiolytic sulfur cycle *Science* **286** 97

[36] Johnson R E, Carlson R W, Cooper J F, Paranicas C, Moore M H and Wong M C 2004 *Jupiter: Planet, Satellites, and Magnetosphere* ed F Bagenal, W McKinnon and T Dowling (Cambridge Univ. Press, Cambridge, UK) p. 485

[37] Hibbitts C A, Klemaszewski J E, McCord T B, Hansen G B and Greeley R 2002 CO_2-rich impact craters on Callisto *J. Geophys. Res.* E **107** 5084

[38] Hibbitts C A, Pappalardo R T, Hansen G B and McCord T B 2003 Carbon dioxide on Ganymede *J. Geophys. Res.* E **108** 5036

[39] Cooper J F, Johnson R E, Mauk B H, Garrett H B and Gehrels N 2001 Energetic ion and electron irradiation of the icy Galilean satellites *Icarus* **149** 133

[40] Gomis O and Strazzulla G 2005 CO_2 production by ion irradiation of H_2O ice on top of carbonaceous materials and its relevance to the Galilean satellites *Icarus* **177** 570

[41] Strazzulla G, Baratta G, Leto G, Gomis O 2007 Hydrate sulfuric acid after sulfur implantation in water ice *Icarus* **192** 623

[42] Dalton J B, Cassidy T, Paranicas C, Shirley J H, Prockter L M and Kamp L W 2013 Exogenic controls on sulfuric acid hydrate production at the surface of Europa Planet *Space Sci.* **77** 45

[43] Lane AL, Nelson R M and Matson D L 1981 Evidence for sulfur implantation in Europa's UV absorption band *Nature* **292** 38

[44] Sack N J, Johnson R E, Boring J W and Baragiola R A 1992 The effect of magnetospheric ion bombardment on the reflectance of Europa's surface *Icarus* **100** 534

[45] Muñoz Caro G M, Dartois E, Boduch P, Rothard H, Domaracka A and Jiménez-Escobar A 2014 High energy ion versus UV irradiation of methanol:ammonia ice - Formation of common organic products *Astron. Astrophys.* **566** A93

[46] Portugal W, Pilling S, Boduch P, Rothard P and Andrade D P P 2014 Radiolysis of amino acids by heavy and energetic cosmic ray analogs in simulated space environments : α-glycine zwitterion form ions *Mon. Not. R. Astron. Soc.* **441** 3209–3225

[47] Allodi M A, Baragiola R A, Baratta G A, Barucci M A, Blake G A, Boduch P, Brucato J R, Contreras C, Cuylle S H, Fulvio D, Gudipati M S, Ioppolo S, Kaňuchová Z, Lignell A, Linnartz H, Palumbo M E, Raut U, Rothard H, Salama F, Savchenko E V, Sciamma-O'Brien E and Strazzulla G 2013 Complementary and emerging techniques for astrophysical ices processed in the laboratory *Space Sci. Rev.* **180** 101-175

[48] Bennett C J, Pirim C and Orlando T M 2013 Space-weathering of Solar System bodies: a laboratory perspective *Chem. Rev.* **113** 9086–9150

Water radiolysis with heavy-ion beams at GANIL. Back to 20 years of investigations.

G. Baldacchino[1]

[1]Physico-Chimie sous Rayonnement, IRAMIS, LIDYL, UMR 3299 CEA-CNRS SIS2M, PC175, CEA Saclay, F-91191 Gif sur Yvette Cedex, France

E-mail: gerard.baldacchino@cea.fr

Abstract. This article reports a short history of the studies carried out in the domain of radiation chemistry with the swift heavy ions of GANIL (Grand Accélérateur National d'Ions Lourds) during the last 20 years. The originality of these studies lays on two main aspects which were never been investigated simultaneously before: 1) the energy of the GANIL ions provides an energy deposition in small samples (few millimeters) that can be considered as almost constant; this allows studies on LET-effect on radiolytic yields, 2) production of pulses as short as a few nanoseconds made available the access to pulse radiolysis method and the study of transient chemical species such as hydrated electron, hydroxyl radical and superoxide. Future is now focus on high temperature effects on water radiolysis for which, again, nothing exist but only simulations and speculations.

1. Context - A brief history of the heavy ion water radiolysis studies

As an introduction of the radiation chemistry investigations initiated at GANIL, we must depict the crucial issues in this domain. Actually the interaction of energetic heavy ion with the liquid water is found in many situations in nuclear industry (PWR, BWR reactors, nuclear waste storage and reprocessing), in hadrontherapy (carbon ions, protons), in cosmic rays involved in future space travel and interstellar chemistry [1]. This particular interaction is closely related to the Linear Energy Transfer (LET) value defined as: $LET = \left(dE/dx\right)_{elec}$ [2, 3].

To depict the effects of LET on the radiolysis of liquid water the use of heavy ion beams of various energies, charges and types are necessary to determine the spatial distribution of the deposited energy, the fate of the deposited energy (excitation, ionization, multi-ionization, etc.) and the time-dependence of the diffusion and reaction processes. For a long time, since the 60's, the physical-chemistry occurring in the ionisation tracks of heavy ions has been described in a general way [2, 4]. The current knowledge in radiation chemistry with high LET particles, and it was many times verified, is that radiolytic yields of the molecular products (H_2 and H_2O_2) increase with the LET whilst those for the radical species ($e_{aq}^{-\bullet}$, $H^\bullet$, $OH^\bullet$) decrease. This general tendency is a result of a more efficient recombination of the radical species to form molecular species in the tracks since the density of the ionisation events is greater with high LET particles. It can be considered as a local dose rate effect. These aspects have been deduced mainly from the studies involving the chemical scavenging method [5-8] which is based on concentration measurements of a final stable product of reactions. In this method the time dependence of the radiolytic yields is determined from the concentration and the rate constant of the reaction between the scavenger and the radical species. This method becomes less accurate at early time after the ionisation process because it needs very high concentrations of scavenger that can also suffer from the direct effect of the ionization.

Pulse radiolysis technique was sporadically used with energetic ions since it depends on the characteristics of accelerators almost essentially devoted to physics investigations. Since the 80's

Burns *et al.* [9-11] have performed some experiments with particles having short ranges in sample and effects on yields of average values of LET were studied. Those were reference experiments for a long time because the nanosecond time resolution has been reached.
In 90's, new hopes in new machines delivered high energy particles. Then it was interesting to compare ions and other new chemical systems for various LET. Some experiments were performed with microsecond time resolution [12-14].
GANIL could provide high energy heavy ions ranging a few millimetres in liquid samples. Its beam control allowed the pulsing of the beam at high rate with a pulse selection. Considering that intrinsic pulse have duration of 1-2 ns, pulse radiolysis studies became evident. Figure 1 shows one of the first setup used at GANIL (IRABAT irradiation port). Many details of this experiments are collected in a recent short review on pulse radiolysis with heavy ions [15]. Actually it collects the technical aspects of the pulse radiolysis method and the specificities by using with heavy ions. Here one can give a short list of publications based on these experiments [16-25]. This method has also been applied to another installation with less energetic ions in Japan [26].

Figure 1. Setup for pulse radiolysis at GANIL (Experiment n°P492 on 19th of June 1999). Center of the picture: a vertical optical cell made of fused silica is irradiated with horizontal beam downstream the output "nose" of the machine made of a thin titanium foil.

Other water radiolysis studies have been carried out at GANIL without using the pulse structure of the beam [27, 28]. It concerned the production of stable products such as hydrogen peroxide (H_2O_2) or molecular hydrogen (H_2). In this short review we will focus essentially on the transient species studied by pulse radiolysis technique coupled to the high-energy ions at GANIL during the last 20 years.

2. Review of the main results by pulse radiolysis

2.1. Ion beam characteristics

The experiment setup used for pulse radiolysis with heavy ions is similar to those used in pulse radiolysis with electron beams or flash photolysis [29]. In principle the method consists of a pulsed source of high energy radiation coming from a particle accelerator and a synchronized acquisition of the transmitted signal. The detection used is commonly absorption spectroscopy, except for an experiment which has used chemiluminescence [20, 21], by using a spectral light of a sufficient intensity at a wavelength chosen to obtain a reasonable sensitivity. An absorbance limit of 10^{-4} has been reached to detect hydrated electron [17]. One special characteristic of the setup is the thickness of the irradiation window and the depth of the cell that are dependent on the ion energy used for the experiment and whether the experimentalist wants to analyze a track segment or a mean effect along the total penetration of the ion in water. Calculations with SRIM program must be achieved to determine its thickness [30].
With GANIL cyclotron, carbon ions with energy of 95 MeV/nucleon (i.e. an energy of 1.1 GeV) can be delivered with an intensity of a few μA. With a cell having a thickness of 0.5 mm of the entrance window, the track segment directly analyzed in the first millimetre after the window has a LET value of 30±1 eV/nm [18]. The main other ions used in the pulse radiolysis experiments was $^{36}Ar^{18+}$ of

95 MeV/nucleon (i.e. 3 GeV) for a LET value of 280 eV/nm. Oxygen, Sulphur and Krypton ions were also used to vary LET.

The essential determinations were the radiolytic yield (G-value) as a function of time, typically in the ionization track and in the expansion period of time where the chemistry is strongly not classical since species are not homogeneously dispersed. We now present some important facts relevant to the implication of the main radicals in water radiolysis.

2.2. Hydrated electron

Hydrated electron ($e_{aq}^{-\bullet}$) has been extensively studied at low LET (mainly with electron beams) and its spectrum centred at 720 nm makes it easy to detect in visible-near IR domain [31]. Its high absorption coefficient allows its detection at very low concentration levels. This is an essential species in water radiolysis since it is the main precursor of H_2. It is also a good candidate to probe the track structure at early time after the ion passage.

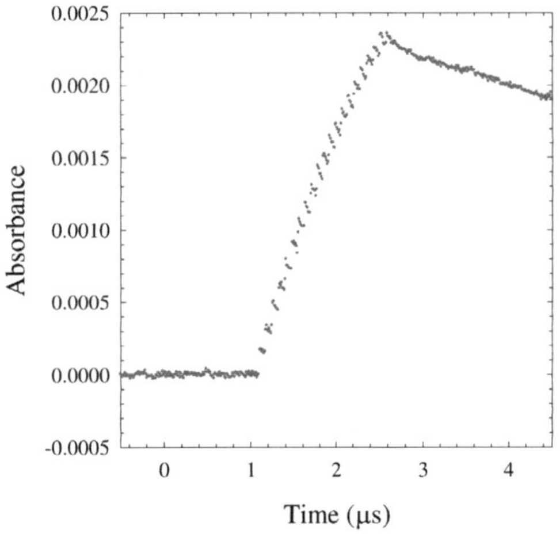

Figure 2. Real time formation of hydrated electron produced by a burst of 20 pulses $^{16}O^{8+}$ of 95MeV/nucleon having a LET of 50 keV/mm. Detection was performed at 670 nm in deaerated water at pH 12 in presence of formate at 0.01M. In these chemical conditions the hydrated electron has the longest life time because it cannot react with H^+ and $OH^\bullet$.

Figure 2 shows how the formation of hydrated electron was detected by using the time structure of GANIL beam. With enough intensity resolution and averaging of enough oscilloscope traces (about 1 million!) one can see the contribution of each ns pulse: the increasing absorbance is saw-teeth shaped. Since the recombination kinetics of hydrated electron is not fast enough due to the chemical conditions depicted in the figure legend, absorption signal is added to the previous one during a macro-pulse of almost 2 μs of duration. Single pulses are separated by a period of 70 ns which corresponds to the radiofrequency (14 MHz) of the cyclotron. Giving a *G*-value from only one single pulse of ions remained difficult because the absorption measurement was about 10-4 which was a real challenge. That means the concentrations were lower than 10^{-7}M. Moreover the dose must have been measured with a high accuracy. After the pulse, the concentration begins a slow decay. This evolution corresponding to the non-homogeneous chemistry is well depicted in the articles [17, 18]. Monte Carlo simulations [32, 33] reproduced the time dependence as far as the experimental *G*-value was suspected to be 10% too high at earliest times [34]. A better accuracy in the dosimetry was needed.

2.3. Hydroxyl radical

The hydroxyl radical is detectable only in deep UV (between 200 and 290 nm with a maximum around 220 nm) with a low absorption coefficient (about 600 $M^{-1}cm^{-1}$ at 210 nm). It is mainly detected by using scavengers. By using pulse radiolysis and the thiocyanate anion for which the reaction mechanism is now more controlled [35] $OH^\bullet$ radicals can be specifically scavenged to give $(SCN)_2^{\bullet-}$ radical which is more convenient to detect at the μs stage in the visible range with a relatively high extinction coefficient (7600 $M^{-1}cm^{-1}$ at 475 nm). By this way one can observe absorption such as in figure 3. This is not a direct detection of $OH^\bullet$ but it gives reliable results if the concentration of thiocyanate is not too high ($<10^{-2}$M) [16] It was showed the scavenger molecule can be subject to multiple possible reactions in the track of heavy ions even by using other scavenger species like the bromide anion.

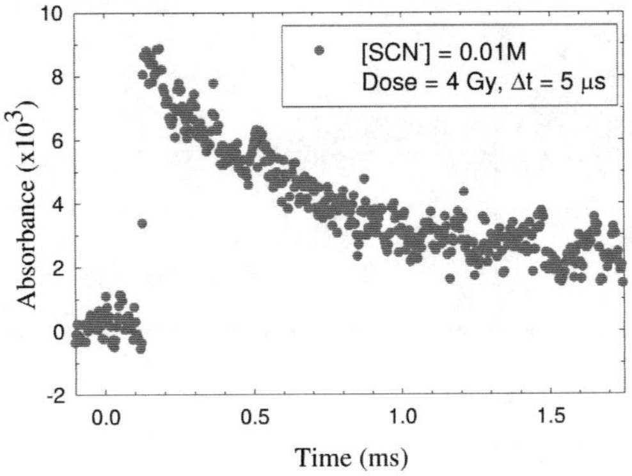

Figure 3. Formation and decrease of the absorption of $(SCN)_2^{\bullet -}$ at 475 nm when using a 5 μs pulse of $^{12}C^{6+}$ of 1 GeV. This was a chemical condition of aerated solution containing 10^{-2} M of KSCN.

Another chemical system has been tested to detect hydroxyl radical (for very high LET value only) by using a light emitting chemical process. This was also a scavenging method which was associated to the pulsed heavy ion beam [19, 21]. In the chemical mechanism of luminol, the luminol molecule must react with hydroxyl and superoxide to form a molecule that emits fluorescence light at around 420 nm. At high LET value, the yield of $OH^{\bullet}$ becomes lower than the yield of $O_2^{\bullet -}$. Therefore, in this condition $OH^{\bullet}$ yield is a limiting value. By this way, at a LET value of 280 eV/nm with Ar^{18+} ions, $OH^{\bullet}$ yield was found to be 2.2×10^{-8} mol/J which was in good agreement with the literature. Due to the concentration of luminol (10^{-3}M), this yield is a primary yield, picture of an homogeneous solution in the time-range 300 ns-1 μs after the energy deposition stage[21]. That method has proved that one could detect very small concentrations of hydroxyl radical by using light emitting system that is known to be more sensitive than absorption spectroscopy [19].

2.4. Superoxide radical

The superoxide radical ($HO_2^{\bullet}/O_2^{\bullet -}$) is a special case in water radiolysis because its radiolytic yield increases with LET value. That is contrary to the recombination rule in dense ionization tracks. As stated before, the general tendency is the radical recombination increases the production of molecular species (H_2 and H_2O_2). The low reactivity of this radical in pure water essentially due to its dismutation (Reaction 1) allows it to escape the track and survives during the dense recombinations in the non-homogeneous stage.

$$HO_2^{\bullet} + O_2^{\bullet -} \xrightarrow{H_2O} H_2O_2 + O_2 + OH^- \qquad k = 9.7 \times 10^7\ M^{-1}s^{-1} \qquad (1)$$

Figure 4 gives a kinetic of formation and dismutation of superoxide under pulse heavy ion beam. To detect the superoxide radical one must know this species does not absorb too much. Its absorption coefficient at 245 nm does not exceed 2500 $M^{-1}cm^{-1}$. Associated to small concentrations, that makes this species difficult to detect. Moreover there is no scavenger available for this species. The only convenience is its lifetime, more than ms. Slow kinetics are of course more convenient to record on a large input impedance of an oscilloscope. By this way, its radiolytic yield has been determined with ions such as Argon and Sulfur. The results have been unique and were simulated with Monte Carlo codes. Models including multiple-ionizations processes have shown they could reproduce the experimental values [36, 37].

Journal of Physics: Conference Series **629** (2015) 012009 doi:10.1088/1742-6596/629/1/012009

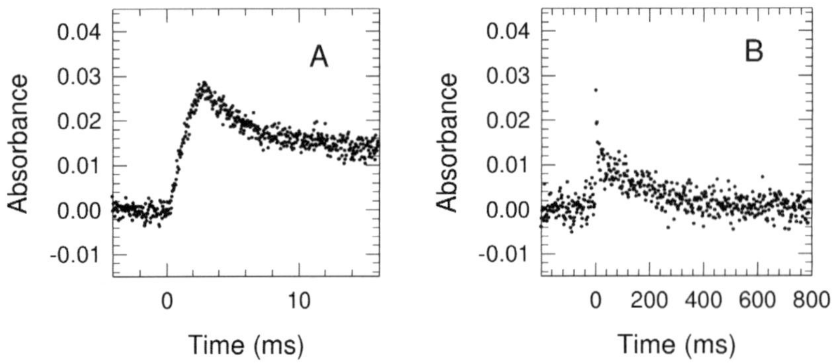

Figure 4. Formation and dismutation of superoxide after 2 ms pulse of Ar^{18+} of 3 GeV. Dose was $9.2x10^8$ ions/s (500 Gy) and solution was deaerated water. The detection wavelength was 260 nm.

3. Future

3.1. Near future: extreme conditions of temperature at GANIL

Higher temperatures than ambient is now crucial for managing PWR water chemistry. Recent database focused on the lack of data for high LET particles and high temperature. These last years, our team has performed original experiment by using a special optical cell designed especially for heavy ions. The challenge is easy to conceive because heavy ions stop on a very short range and pressure due to temperature needs a strong optical cell, and therefore a sufficient thickness of material. This dilemma was overcome with a special designed window in Inconel 718 of 0.4 mm of thickness. Figure 5 shows the final optical cell containing 2 sapphire windows. Details were given in recent thesis report and article [38, 39].

Figure 5. High temperature optical flow cell dedicated to Heavy ion irradiations. Imagine the ion beam come from the right side; ions enter in the hole of 3mm diameter to access the internal interaction chamber of 1.5mm of diameter. The main body is made of Inconel 718 in a 51 mm long and 51 mm diameter cylinder. Heaters (with their own thermocouples) and tubes are also going to the left. The picture does not include the main thermocouple and thermal protection made of silica wool.

We expect to deliver soon results concerning the radiolytic yields of $HO^{\bullet}$ radicals as a function of temperature up to supercritical water conditions with a range of LET values from 30 to 250 eV/nm as it was already published for a limited range of temperature [39].

3.2. Future: Increasing time resolution

In the near future the ion-beam investigations will probably focus on earliest processes in the track of heavy ions. Accelerators must deliver shorter pulses than currently available and the detection must be more sensitive and highly time-resolved. The project of picosecond pulse radiolysis research with heavy ions is possible because new designs of accelerators for proton and heavier ions swift by laser driven accelerator have already started [40-43].

Nevertheless orientations of GANIL with SPIRAL project may not concern at all radiation chemistry in the future in terms of shorter and shorter ion pulse durations[44]. Other installations based on high intensity femtosecond laser are much in phase with the radiation chemistry issues.

Acknowledgements
The author thanks Dr. B. Hickel for his brilliant ideas during the first experiments at GANIL, for his management of the first studies in France with pulsed heavy ions. The experiments were performed at GANIL in collaboration with our first local contact Dr S. Bouffard and these last years with Dr E. Balanzat. We thank their availability and their great knowledge of the beam control. We acknowledge also the GANIL staff and the colleagues in the Laboratoire de Radiolyse at Saclay for their great help during the runs: Dr S. Pin, G. Vigneron, Dr J. P. Renault, Dr S. Le Caër, Dr S. Pommeret and Dr J.-C. Mialocq. The first experiments were also performed in collaboration with Pr M. Gardès-Albert (Paris Descartes University) and Pr S. Deycard (Caen University). Chemistry at GANIL was a big challenge at this time. They had the privilege to start the first irradiations of solutions in the 90's. Thanks for their help and discussions.

References

[1] Hatano Y, Katsumura Y, Mozumder A (eds) 2011 *Charged Particle and Photon Interactions with Matter. Recent Advances, Applications, and Interfaces* (Boca Raton : CRC Press)

[2] Baldacchino G and Katsumura Y 2010 Chemical processes in heavy ions track *Recent Trends in Radiation Chemistry* ed J F Wishart and B S M Rao (Singapore : World Scientific) pp 231-254

[3] Yamashita S, Taguchi M, Baldacchino G and Katsumura Y 2011 Radiation Chemistry of Liquid Water with Heavy Ions: Steady-States and Pulse Radiolysis Studies *Charged Particle and Photon Interactions with Matter. Recent Advances, Applications, and Interfaces* ed Y Hatano *et al* (Boca Raton : CRC Press) pp 325-354

[4] Allen A 1961 *The Radiation Chemistry of Water and Aqueous Solutions* ed D Van Nostrand (Princeton, NJ)

[5] LaVerne J A and Yoshida H 1993 *J. Phys. Chem.* **97** 10720-10724

[6] LaVerne J A and Pimblott S M 1993 *Radiat. Res.* **135** 16-23

[7] LaVerne J A and Pimblott S M 1991 *J. Phys. Chem.* **95** 3196-3206

[8] LaVerne J A 1989 *Radiat. Res.* **118** 201-210

[9] Rice S A, Playford V J, Burns W G and Buxton G V 1982 *J. Phys. E. Sci. Intrum.* **15** 1240-1243

[10] Burns W G, May R, Buxton G V and Wilkinsontough G S 1981 *J. Chem. Soc. Faraday Trans. 1* **77** 1543-1551

[11] Burns W G, May R, Buxton G V and Tough G S 1977 *Faraday Discuss.* **63** 47-54

[12] Chitose N, Katsumura Y, Domae M, Cai Z L, Muroya Y, Murakami T and LaVerne J A 2001 *J. Phys. Chem.* A **105** 4902-4907

[13] Chitose N, Katsumura Y, Domae M, Zuo Z H, Murakami T and LaVerne J A 1999 *J. Phys. Chem. A* **103** 4769-4774

[14] Chitose N, Katsumura Y, Zuo Z H, Domae M, Ishigure K and Murakami T 1997 *J. Chem. Soc. Faraday Trans.* **93** 3939-3944

[15] Baldacchino G 2008 *Radiat. Phys. Chem.* **77** 1218-1223

[16] Baldacchino G, Vigneron G, Renault J P, Le Caër S, Pin S, Mialocq J C, Balanzat E and Bouffard S 2006 *Nuc. Instr. Meth.* B **245** 288-291

[17] Baldacchino G, Vigneron G, Renault J P, Pin S, Abedinzadeh Z, Deycard S, Balanzat E, Bouffard S, Gardes-Albert M, Hickel B and Mialocq J C 2004 *Chem. Phys. Lett.* **385** 66-71

[18] Baldacchino G, Vigneron G, Renault J P, Pin S, Remita S, Abedinzadeh Z, Deycard S, Balanzat E, Bouffard S, Gardes-Albert M, Hickel B and Mialocq J C 2003 *Nuc. Instr. Meth.* B **209** 219-223

[19] Wasselin-Trupin V, Baldacchino G and Hickel B 2001 *Can. J. Physiol. Pharm.* **79** 171-175

[20] Wasselin -Trupin V, Baldacchino G and Hickel B 2001 *Can. J. Physiol. Pharm.* **79** 4

[21] Wasselin-Trupin V, Baldacchino G, Bouffard S, Balanzat E, Gardes-Albert M, Abedinzadeh Z, Jore D, Deycard S and Hickel B 2000 *J. Phys. Chem.* A **104** 8709-8714

[22] Baldacchino G, Trupin V, Bouffard S, Balanzat E, Gardes-Albert M, Abedinzadeh Z, Jore D,

Deycard S and Hickel B 1999 *J. Chim. Phys.* PCB **96** 50-60

[23] Baldacchino G, Le Parc D, Hickel B, Gardes-Albert M, Abedinzadeh Z, Jore D, Deycard S, Bouffard S, Mouton V and Balanzat E 1998 *Radiat. Res.* **149** 128-133

[24] Baldacchino G, Bouffard S, Balanzat E, Gardes-Albert M, Abedinzadeh Z, Jore D, Deycard S and Hickel B 1998 *Nucl. Instrum. Meth. B* **146** 528-532

[25] Baldacchino G, Bouffard S, Balanzat E, Mouton V, GardesAlbert M, Abedinzadeh Z, Jore D, Deycard S, LeParc D and Hickel B 1997 *J. Chim. Phys. PCB* **94** 200-204

[26] Taguchi M, Baldacchino G, Kurashima S, Kimura A, Sugo Y, Katsumura Y and Hirota K 2009 *Radiat. Phys. Chem.* **78** 1169-1174

[27] Wasselin-Trupin V, Baldacchino G, Bouffard S and Hickel B 2002 *Radiat. Phys. Chem.* **65** 53-61

[28] GardesAlbert M, Jore D, Abedinzadeh Z, Rouscilles A, Deycard D and Bouffard S 1996 *J. Chim. Phys.* PCB **93** 103-110

[29] Tabata Y (ed) 1991 *Pulse Radiolysis* (Boca Raton : CRC Press)

[30] Ziegler J F, Ziegler M D and Biersack J P 2010 *Nucl. Instrum. Meth.* B **268** 1818-1823

[31] Spinks J W T and Woods R J 1990 *An Introduction to Radiation Chemitry* ed J Wiley (New York)

[32] Frongillo Y, Goulet T, Fraser M J, Cobut V, Patau J P and Jay-Gerin J P 1998 *Radiat. Phys. Chem.* **51** 245-254

[33] Cobut V, Frongillo Y, Patau J P, Goulet T, Fraser M J and Jay-Gerin J P 1998 *Radiat. Phys. Chem.* **51** 229-243

[34] LaVerne J A, Stefanic I and Pimblott S M 2005 *J. Phys. Chem.* A **109** 9393-9401

[35] Milosavljevic B H and LaVerne J A 2005 *J. Phys. Chem.* A **109** 165-168

[36] Gervais B, Beuve M, Olivera G H, Galassi M E and Rivarola R D 2005 *Chem. Phys. Lett.* **410** 330-334

[37] Gervais B, Beuve M, Olivera G H and Galassi M E 2006 *Radiat. Phys. Chem.* **75** 493-513

[38] Saffré D 2011 Radiolyse de l'eau dans des conditions extrêmes de température et de TEL. Capture de HO• par les ions Br- *Ecole doctorale de Chimie de Paris-Sud – PhD thesis* (Orsay, Université de Paris-Sud) p 260

[39] Balcerzyk A, Boughattas I, Pin S, Balanzat E and Baldacchino G 2014 *Phys. Chem. Chem. Phys.* **16** 23975-23984

[40] Cheriaux G, Giambruno F, Freneaux A, Leconte F, Ramirez L P, Georges P, Druon F, Papadopoulos D N, Pellegrina A, Le Blanc C, Doyen I, Legat L, Boudenne J M, Mennerat G, Audebert P, Mourou G, Mathieu F and Chambaret J P 2012 Light at extreme Intensities 2011 (nov. 2011, Szeged, Hungary) *AIP Conf. Proc.* **1462** 78-83

[41] Floquet V, Ceccotti T, Dufrenoy S D, Bonnaud G, Gremillet L, Monot P and Martin P 2012 *Phys. Plasmas* **19** 094501

[42] Lefebvre E, Gremillet L, Levy A, Nuter R, Antici P, Carrie M, Ceccotti T, Drouin M, Fuchs J, Malka V and Neely D 2010 *New J. Phys.* **12** 045017

[43] Malka V and Mora P 2009 *C. R. Phys.* **10** 106-115

[44] Lewitowicz M 2008 *Nucl. Phys.* A **805** 519-525c

Crystal assisted experiments for multi-disciplinary physics with heavy ion beams at GANIL

Denis Dauvergne
Université de Lyon, Université Lyon 1, CNRS/IN2P3, Institut de Physique Nucléaire de Lyon (UMR 5822), 69622 Villeurbanne, France

E-mail: d.dauvergne@ipnl.in2p3.fr

Abstract. We present a review of the channeling and blocking experiments that have been performed at GANIL during the 30 years of stable beam operation, with the strong support of the multi-disciplinary CIRIL-CIMAP laboratory. These experiments combine atomic physics, solid state physics and nuclear physics.

1. Introduction: Channeling and blocking of swift heavy ions

Briefly, ions travelling in a crystal with a propagation direction close to a major axial or planar direction undergo a series of correlated collisions with aligned atoms, which leads to a non-uniform flux and anisotropies in the angular distribution of the particles inside the crystal [1, 2]. At sufficiently large distances from atomic strings or planes, i.e. in conditions where no nuclear or large angle elastic collisions occur with target nuclei, the continuum potential approximation allows one to describe channeled ions as particles trapped into a 1D or 2D potential well, in planar or axial channeling conditions, respectively. Particles at the crystal entrance are given a transverse energy E_t that depends on the incidence angle ψ_{in} and the transverse position r_{in} :

$$E_t = E{\psi_{in}}^2 + qU(r_{in}).$$

If E_t is small enough, typically $E_t < qU(\rho_{th})$ where ρ_{th} is the thermal vibration amplitude of the crystal atoms, then one can consider that ions are channeled, and keep their transverse energy throughout the crystal in first approximation. The above condition was expressed in terms of a critical angle by Lindhard [3]. In the case of axial channeling, the incidence angle of the beam relative to the axis should be smaller than $\psi_c = (2Z_1Z_2e^2/Ed)^{1/2}$.

The blocking effect corresponds to a reduction of the probability, for particles emitted close to atomic sites of the crystal, to emerge in a direction parallel to crystallographic ones. *Stricto sensu*, channeling and blocking are reciprocal effects, and they are governed by the reversibility rule: both correspond to a redistribution of the ion flux inside the crystal. Channeling leads to a reduction of close collisions with atomic cores when a parallel beam is aligned along a crystallographic direction, whereas blocking is a reduction of the flux parallel to a crystallographic direction for ions emitted at close distance from an atomic core. The same characteristic critical angle governs both processes. A particular application of blocking is the measurement of nuclear lifetimes at the attosecond scale by measuring transverse nuclear recoils with respect to atomic strings or planes.

2. Material and method

The GANIL accelerator is a unique facility to provide high intensity, low emittance beams of heavy ions at a few tens of MeV per nucleon. Ions produced in an ECR source are injected in a first cyclotron C0, and then in a first separated-segment cyclotron CSS1, after which they reach 5-6 MeV/u for the heaviest species, then stripped in a thin foil before injection into the second CSS2 cyclotron, reaching 24 MeV/u for U beams. High energy ions emerging from CSS2 have typically the most probable charge state of the medium energy after the stripper foil, i.e. they are hardly fully stripped for Z>20. Typical intensities of 10-100 nA are available within an emittance value of 15π mm×mrad. The maximum intensity depends on the nature of the ion species (rare gas ions are produced much more easily than metallic ions for instance). Further stripping and energy degrading may be performed by inserting a target of suitable thickness and composition at the exit of CSS2. In order to realize detailed channeling experiments, strong reduction of the emittance has to be performed. This is achieved upstream and downstream of the GANIL's alpha spectrometer, by means of two series of three slits and dedicated quadrupoles. The momentum dispersion of the beam may be reduced at the dispersive plane of the alpha spectrometer by means of another slit. The whole alpha spectrometer is achromatic. After acceleration and tailoring of the beam with adequate emittance (of the order of 0.1 π mm×mrad in both directions), energy spread, charge state and intensity, the ion beam is guided and distributed into the various experimental areas.

The first experiments took place in the LISE beam line, for which beam optics was optimized for channeling. It comprises a dipole magnet with a dispersion of 1.8 cm/%dp/p at the focal plane, and a vertical-slit collimation to perform charge- or energy-selection downstream of this dipole. A multiwire gas proportional chamber (MWPC) could be used for transmitted beam profile acquisition, and therefore to energy loss and charge state distribution measurements. Later, the high-resolution SPEG was used (8 cm/%dp/p at the image plane) equipped with position sensitive MWPC with event-by-event detection capability.

3. Experimental program

One of the particularities of the ions accelerated at GANIL is that heavy ions are not fully stripped, and charge exchange plays an important role in particle matter-interaction. Another is that high electronic energy loss regimes are achieved with swift heavy ions. This motivated the implementation of atomic and solid-state physics research programs, hosted by the CIRIL laboratory.

As soon as GANIL was built, a large manifestation of interest took place among several multidisciplinary laboratories for setting up channeling experiments. These experiments were hosted by the CIRIL (now CIMAP). The initial collaboration gathered solid state physicists from the Groupe de Physique du Solide (Paris, later renamed as INSP), Laboratoire des Solides Irradiés (Fontenay aux Roses), atomic and nuclear physicists from Institut de Physique Nucléaire de Lyon, Centre de Recherche Nucléaire (Strasbourg), and a bit later Centre d'Etudes Nucléaires de Bordeaux. Beam experts from GANIL were also contributing. The IPNL, INSP and CIRIL groups have been continuously working on this program during more than 20 years.

The first GANIL-channeling publication, by Camille Cohen et al. [4] dealt with 60 MeV/u Ar^{18+} ions channeled in a Ge crystal. Products of nuclear reactions (gamma, neutrons) were used to characterize the extinction of close collisions, and the observed values of critical angles and extinction yields were found – as expected – in agreement with extrapolated values from lower energy experiments.

3.1. Charge exchange, energy loss and related surface-emission studies

In amorphous solids or random crystal orientation condition, Nuclear Impact Ionization (NII) and non-radiative electron capture, also called Mechanical Electron Capture (MEC), are the dominant processes in the charge exchange by heavy ions in the high-energy domain covered by GANIL. In contrast, charge exchange by channeled ions is dominated by ion-electron interaction processes that do not involve nuclear recoil during close collisions. In that way, the crystal target can be considered as a dense and non-homogenous electron target. The various processes that have been studied at GANIL are presented in figure 1, with a representation of the electronic transitions in the ion energy frame.

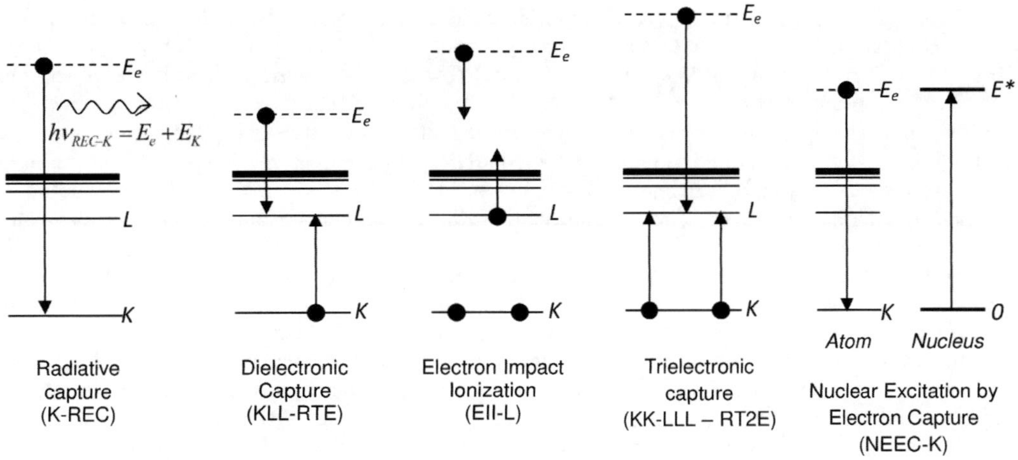

Figure 1: Charge exchange resulting from ion-electron interaction that have been studied in channeling conditions .

The first charge exchange study dealt with Radiative Electron Capture (REC) by hydrogen-like xenon ions at 25 MeV/u, which lead to a measurement of K- and L-REC cross sections [5]. As the adiabaticity parameter $\eta_K = (v_p/Z_p v_0)^2$ (where Z_p is the ion atomic number, v_p the ion velocity, v_0 the Bohr velocity) is much smaller than 1, this means that the K-shell is filled at charge state equilibrium. Thus K-REC cannot be observed in random crystal orientation, because non-radiative electron capture, also called Mechanical Electron Capture (MEC), followed by radiative decay, rapidly fills the K-shell vacancy. Therefore one has to use the property of the extinction of close-collision induced-charge exchange to observe REC, which results from pure ion-electron interaction. This was a first step toward electron capture into inner-shell at out-of-equilibrium conditions. At the same experiment, several incident charge states were used, and the incident charge state freezing in axial condition could be studied in detail [6].

A detailed study of energy loss and Electron Impact Ionization (EII) was reported by A. L'Hoir et al.[7]. In this experiment, 27 MeV/u Xe^{35+} ions were used as projectile into a 21 µm thick silicon crystal aligned along the <110> axial direction. In this case, the M-shell ionization probability is very high, even for EII, and thus EII rates could be used as a probe for the local electron density evaluation along the ion path in the crystal. Combining these studies with energy loss measurements led to a detailed study of local and non-local parts of the energy loss as a function of transverse energy. One of the conclusions was that, at such not-so-high velocities, the energy loss of low –transverse energy ions is due only to the valence electron gas, however it is almost independent on the local electron density. The extinction of the contribution of silicon L-electrons to energy loss was observed to be more dependent on the local electron density.

In collaboration with the IKF-Frankfurt, electron emission spectroscopy at the crystal exit was performed in 1990, using 27 MeV/u Xe^{37+} incident ions on a 21 µm thick crystal. The results were only partly published in [8]. An extinction of the binary encounter electron yield (at velocity close to 2 v_p at 0°) in axial crystal orientation was found equal to the ratio of valence to total electron density in silicon, showing that this yield depends on the local electron density.

The same year, an experiment dedicated to the dielectronic capture, also called Resonant Transfer and Excitation- RTE in the case of bound target electrons - took place in the LISE beam line. Thanks to a particular effort from the accelerator team, a full resonance was described, with energy variation of a Xe^{52+} ion beam from 42.3 to 33.8 MeV/u, while keeping constant the beam characteristics at the target point. The resonance was observed by means of K_α radiative decay and by charge state distributions as a function of energy loss measured after charge and energy analysis. The shape of the resonance was compared to calculated Compton profiles [9]. This was the last of four experiments performed with the same 21µm thick silicon crystal. Radiation damage is quite reduced in channeling conditions.

The following experiment – during which the atomic physics group of the GPS-Paris joined the collaboration - was again devoted to REC, but this time in conditions where REC would be observable both in channeling and random conditions, i.e. at η value greater than 1. This was achieved with krypton ions at 60 MeV/u into a 37 µm thick silicon crystal. Coincidences between X-rays and transmitted ions with given charge state were used to sign electron capture or ionization. Using high statistics and high resolution X-ray spectra, the shape of the K- and L-REC lines were used to infer impact parameter dependent Compton profiles from the various electronic levels of silicon (valence and core electrons) [10]. A by-product of this experiment was the measurement of intrashell excitation which depopulates metastable n=2 states of H-like or He-like ions, even in channeling conditions [11].

Then followed an attempt to observe Trielectronic Capture or RT2E (single electron capture accompanied with double excitation of the ion), namely KK-LLL, by means of triple coincidence between a single capture event and two K□ photon decay. The first attempt, using Nb^{39+} ions in LISE, was interrupted by a fire in the injection cyclotron. The final attempt was the first channeling experiment performed in the high-resolution spectrometer SPEG, using He-like Kr ions, with GANIL researchers. An upper limit for the cross section of this effect was obtained [12].

The next experiment in SPEG took place in 2000. The objective was twofold: to study the dependence of electron emission from both entrance and exit surfaces with the transverse energy of incident Pb^{56+} ions at 29 MeV/u, and make detailed study of energy loss as a function of charge exchange in a very thin crystal (1.1 µm thickness). Backward electron yields were found to be correlated to the impact parameter at the crystal entrance, whereas forward yields were found less correlated to the transverse energy [13]. Another result from this experiment was the observation of the superdensity effect, which leads to ions with higher charge state and higher energy loss than obtainable in random conditions, due to series of close collisions at small impact parameter that enhance the probability of multi-step ionization, and prevent decay into deeply bound states after capture. The corresponding excess in energy loss during close collisions is not compensated by lower electron density sampled at large distance from the strings inside such a thin crystal [14].

In collaboration with CEA-DAM and CENBG-Bordeaux, an experiment was set in the LISE beam line to observe NEEC (the time-reverse process of internal conversion) by $^{57}Fe^{26+}$. The aim was to excite the 14.4 keV nuclear transition by resonant capture into the K-shell at 9.4 MeV/u, and to detect the signature of the in-flight decay of the excited level by internal conversion, in the 6 meter straight section between the two dipoles of LISE. The vacuum conditions in the beam line prevented us to reach the sensitivity required to observe the microbarn cross-section-NEEC. So far this process has still never been evidenced, and channeling remains a competitive technique to observe it, since an aligned crystal of a few microns is equivalent to a pure electron target of $\sim 10^{20}$ e/cm^2.

The last experiment performed by the collaboration in 2006 (with IPN-Orsay collaborators) consisted in the measurement of ionic sputtering by the surface of a germanium crystal. The energy loss of an incident ion at the entrance of an axially aligned crystal varies over one order of magnitude as a function of transverse energy. This is expected to influence strongly the ion emission yield. An ionic mass spectrometer was set at the entrance face of a thin germanium crystal in SPEG, and secondary ions were detected as a function of the charge state of Pb incident ions at 29 and 5.6 MeV/u. With the actual experimental conditions (secondary vacuum), mainly organic molecular ions from the surface were detected. However, we could observe that emission from amorphous compounds covering the surface is strongly correlated to the energy loss below the surface, where channeling acts for differentiating this particular feature of energy deposition [15].

3.2. Resonant Coherent Excitation
Resonant Coherent Excitation (RCE) makes use of the lattice periodicity to induce internal excitation of the projectile. In a collaboration between IPNL-Lyon, GPS-Paris, CIRIL-Caen and CENBG-Bordeaux, a feasibility study to observe nuclear RCE of a low-energy transition of a metastable state of ^{45}Ti (3.45 keV) was undertaken in 1994. However, since the candidate nucleus has to be produced by means of fragmentation, the beam intensities achievable at GANIL at that time could not allow one to observe this resonance [16].

About the same period, we certainly missed the opportunity to collaborate with Sheldon Datz, who proposed to the members of the collaboration to use planar channeling for varying the excitation frequency during atomic RCE experiments. Using this method, and with the very low momentum dispersion and low emittance achievable with the GANIL beams, precision experiments would certainly have been successful. This way of studying atomic RCE is still employed by the Japanese group of Azuma et al. [17].

3.3. Nuclear fission times
Nuclear physics applications of ion-crystal interaction have been studied experimentally at GANIL. In addition to the search for NEEC and nuclear RCE already discussed above, nuclear lifetime measurements by the blocking technique were performed. The first experiment was done by J. Gomez Del Campo et al. [18] who measured long-lived projectile-like fragments during the bombardment of a germanium crystal by argon ions.

Later, a collaboration between GANIL, IPN-Lyon, INSP-PARIS, CEA-Saclay, IPN-Orsay and CIRIL-Caen performed a series of experiments to measure fission times of uranium and lead during collisions with a silicon crystal at high energy. The use of inverse kinematics allows one to detect both fission fragments. The excitation energy was measured by the ORION neutron detector. In the case of uranium fission, strong variation of the average fission time was found as a function of excitation energy [19]. As the effect corresponds mostly to a uniform filling of blocking dips as the excitation energy decreases, the interpretation by means of simulations relies on an increase of a fraction of long fission times for which no blocking is observed (i.e. when recoil nuclei travel typically more than half of the transverse cell before fission) [1]. For ^{208}Pb ions that are much less fissile, a trend was found for the existence of non-short lifetimes ($t > 10^{-18}$ s typically) when the excitation energy was the smallest and for fissionning nuclei $Z_1+Z_2 \leq 82$ (sum of the two fragment atomic numbers) [20]. Note that, in this experiment, the comparative analysis of fission times and elastic scattering recoils (used as reference for instantaneous events) was made for blocking along major planes, one of which being almost coplanar with the beam direction. Therefore, the time needed for a nucleus to recoil at distances larger than the thermal vibrations is increased by one order of magnitude with respect to other planes. The observed evidence for a decrease of the blocking effect along this plane was a clear signature of fission times that are even larger than 10^{-17}s [21].

These fission experiments were followed by the study of heavy compound nuclei ($Z_{proj} + Z_{crystal} \geq 114$) to infer the restauration of Coulomb barrier close to the expected stability-island of super-heavy nuclei, using low energy uranium and lead nuclei (CSS1 solo), and the INDRA multi-detector to detect all fission products, including light fragments. The results showed evidence for the existence of long lifetimes ($t > 10^{-18}$s) for Z= 120 and 124, but not for Z=114 [22]. However, as shown in the quoted article, not all blocking dips could be reproduced by simulations because of the experimental resolution, and further investigations would be needed to show unambiguously that no long fission times are observed for other species.

3.4. Cooling and heating
In the late 90's – early 2000's, W. Assmann et al. have shown evidence for the breaking of the reversibility rule with low energy heavy ions, i.e. non uniform flux was found at emergence for uniform incidence with angular divergences large compared to critical channeling angles. Either cooling or heating of the transverse energy could be observed, depending on the ion energy, the ion and crystal atomic numbers, and the considered crystallographic direction. A large amount of experimental data was accumulated, among them some at GANIL in collaboration with CIMAP, and the interpretation was given in terms of impact parameter charge exchange and its influence on the conservation of transverse energy [23].

3.5. Ion guiding with micro-capillary
It has been shown that low energy charged ions can be guided through insulating pores or capillaries, by means of a dynamic charge patch formation at places where the incident particles hit the walls inside the capillaries [24]. Since then many groups explored this phenomenon that is closely related to channeling, although the guiding potential is not due to the static screened atomic potential. Indeed similar scaling laws were found as in channeling to describe the motion of guided ions. A collaboration between CIMAP, Riken and Kalamazoo studied these properties with glass capillaries of microscopic dimensions, using the low-energy beam lines of GANIL and multiply charged heavy ions [25].

4. Concluding remarks
The channeling and blocking phenomena offer unique techniques to perform multi-disciplinary studies with fast heavy ion beams, with applications both in nuclear physics (nuclear lifetime measurements, observation of rare Coulomb excitation modes like nuclear RCE or NEEC), atomic physics (detailed ion-electron interaction related charge-exchange processes like REC, RTE, EII), ion-solid interaction (impact parameter dependent energy loss, surface emission studies, surface and bulk damage). The high quality of stable beams and the high-resolution spectrometers and detectors available at GANIL, offer excellent opportunities to perform such experiments. The studies performed at GANIL are quite complementary to those performed at GSI, where higher energy beams are available, but also higher charge-states up to the heaviest ions at low energy after stripping and deceleration (see e.g. [26]). The study of charge exchange processes may have reached some limits, as well as nuclear physics studies, beyond which experiments become quite complex and require very long beam time. In the next future, application of micro-beams delivered through micro-capillaries at low energy may find applications in lithography and precise biological material irradiation. The medium- and low- energy highly charged ions of GANIL that correspond to the maximum of dose deposition in solids are still promising for new physical results, by using the selective energy loss associated to non-random trajectories inside crystal channels as performed in a SiO_2 crystal [27]. In particular, surface and bulk damage studies could be performed under ultra-high vacuum. An equipped chamber is now available [28], but still a goniometer needs to be implemented in this UHV setup.

Acknowledgements
All these experiments would not have been done without a large scientific collaboration based on the different French groups quoted in the text. It should be noted that external collaborations with Chinese

researchers (The Hong Kong University of Science and Technology), German researchers (GSI, Darmstadt) and Japanese researchers (University of Tokyo), were very fruitful. Moreover these experiments would never have been possible without the deep help of the staffs of GANIL, especially R. Anne in charge of the LISE line, and of CIMAP. We cannot forget three of our colleagues who have just passed away: C. Cohen, one of the pillar researchers of all these experiments, J. Dural and J.F. Libin who have taken in charge all technical aspects of the experiments, either in LISE or SPEG areas respectively. Last, the core of the collaboration succeeded in working together during more than 25 years. Such a remarkable longevity would not have been achieved without rigorous scientific motivation and exchanges, mutual trust among all partners, and a friendly climate. The author thanks Marcel Toulemonde, Alain L'Hoir and Jean-Claude Poizat for their contribution.

References

[1] Cohen C, Dauvergne D 2004 High energy ion channeling. *Nucl. Instrum Meth.* B **225** 40-71
[2] Gemmell D S 1974 Channeling and related effects in the motion of charged particles through crystals *Rev. Mod. Phys.* **46** 129-227
[3] Lindhard J 1965 Influence of crystal lattice on motion of energetic charged particles *Mat. Fys. Medd. Dan. Vid. Selsk.* **34** 1
[4] Cohen C, Dural J, Gaillard M J, Genre R, Grob J J, Hage-Ali M, Kirsch R, L'Hoir A, Mory J, Poizat J C, Quéré Y, Remillieux J, Schmaus D and Toulemonde M 1985 Channelling of 2.4 GeV Ar ions in a germanium crystal *J. Phys.* **46** 1565-9
[5] Andriamonje S, Chevallier M, Cohen C, Dural J, Gaillard M J, Genre R, Hage-Ali M, Kirsch R, L'Hoir A, Mazuy B, Mory J, Moulin J, Poizat J C, Remillieux J, Schmaus D and Toulemonde M 1987 Observation of radiative electron capture into K, L, M shells of 25-MeV/u Xe^{53+} ions channeled in silicon *Phys. Rev. Lett.* **59** 2271-4
[6] Quéré Y, Andriamonje S, Chevallier M, Cohen C, Dural J, Genre R, Girard Y, Groeneveld K O, Kemmler J, Kirsch R, L'Hoir A, Poizat J C, Remillieux J, Schmaus D, Schosnig M and Toulemonde M 1991 GeV xenon ions channeled through a silicon crystal *Rad. Ef.f Def. Sol.* **117** 63-7
[7] L'Hoir A, Andriamonje S, Anne R, De Castro Faria NV, Chevallier M, Cohen C, Dural J, Gaillard M J, Genre R, Hage-Ali M, Kirsch R, Farizon-Mazuy B, Mory J, Moulin J, Poizat J C, Remillieux J, Schmaus D and Toulemonde M 1990 Impact parameter dependence of energy loss and target-electron-induced ionization for 27 MeV/u Xe^{35+} incident ions transmitted in [110] Si channels *Nucl. Instrum Meth.* B **48** 145-55
[8] Andriamonje S, Chevallier M, Cohen C, Dural J, Genre R, Girard Y, Groeneveld K O, Kemmler J, Kirsch R, L'Hoir A, Maier R, Poizat J C, Quéré Y, Remillieux J, Schmaus D and Toulemonde M 1991 Charge exchange processes of high energy heavy ions channeled in crystals *High-Energy Ion-Atom Collisions* Springer Berlin Heidelberg pp 221-34
[9] Andriamonje S, Chevallier M, Cohen C, Cue N, Dauvergne D, Dural J, Genre R, Girard Y, Kirsch R, L'Hoir A, Poizat J C, Quéré Y, Remillieux J, Schmaus D and Toulemonde M 1992 RTE measurement with Xe^{52+} ions channeled in a Si crystal *Phys. Lett.* A **164** 184-90
[10] Andriamonje S, Chevallier M, Cohen C, Cue N, Dauvergne D, Dural J, Fujimoto F, Kirsch R, L'Hoir A, Poizat J C, Quéré Y, Remillieux J, Röhl C, Rothard H, Rozet J P, Schmaus D, Toulemonde M and Vernhet D 1996 K-shell radiative electron capture with bare 60-MeV/u Kr ions channeled in a Si crystal: Experiments and simulations *Phys. Rev.* A **54** 1404-16.
[11] Andriamonje S, Chevallier M, Cohen C, Cue N, Dauvergne D, Dural J, Fujimoto F, Kirsch R, L'Hoir A, Poizat J C, Quéré Y, Remillieux J, Röhl C, Rothard H, Rozet J P, Schmaus D, Toulemonde M and Vernhet D 1996 Influence of intrashell excitation (n = 2) on the

population of mestastable states of H- and He-like krypton ions in channeling conditions *Nucl. Instrum Meth.* B **107** 1-5

[12] Hevallier M, Cohen C, Cue N, Dauvergne D, Dural J, Gangnan P, Kirsch R, L'Hoir A, Lelièvre D, Libin J F, Mokler P H, Poizat J C, Prinz H T, Ramillon J M, Remillieux J, Roussel-Chomaz P, Rozet J P, Sanuy F, Schmaus D, Stephan C, Toulemonde M, Vernhet D and Warczak A 2000 Upper-limit determination of resonant trielectronic recombination cross-section for krypton using crystal channeling *Phys. Rev.* A **61** 022724

[13] Barrue F, Chevallier M, Dauvergne D, Kirsch R, Poizat J-C, Ray C, Adoui L, Cassimi A, Rothard H, Toulemonde M, Cohen C, L'Hoir A, Vernhet D, Demonchy C, Giot L, Mittig W, Pita S, Roussel-Chomaz P and Billebaud A 2004 Electron emission induced by fast heavy ions in a thin silicon crystal *Phys. Rev.* A **70** 032902

[14] L'Hoir A, Adoui L, Barrué F, Billebaud A, Bosch F, Bräuning-Demian A, et al. 2006 Ion slowing down and charge exchange at small impact parameters selected by channeling: superdensity effects *Nucl. Instrum Meth.* B **245** 1-14

[15] L'Hoir A, Koumeir C, Della Negra S, Boduch P, Roussel-Chomaz P, Cassimi A, et al. 2009 Study of ion emission from a germanium crystal surface under impact of fast Pb ions in channeling conditions *Nucl. Instrum Meth.* B **267** 876-80

[16] Andriamonje S, Blank B, Del Moral R, Dufour J-P, Faux L, Fleury A, et al. 1994 Search for the observation of Resonant Coherent Nuclear Excitation of channeled isomeres. Nouv GANIL **49** 15

[17] Nakano Y, Takano Y, Ikeda T, Kanai Y, Suda S, Azuma T, et al. 2013 Resonant coherent excitation of the lithiumlike uranium ion: A scheme for heavy-ion spectroscopy *Phys. Rev.* A **87** 060501

[18] Gomez del Campo J, Barrette J, Dayras RA, Wieleczko JP, Pollacco EC, Saint-Laurent F, et al. 1990 Measurements of time delays for projectile-like fragments in the reaction ^{40}Ar+Ge at 44 MeV/nucleon *Phys. Rev.* C **41** 139-46

[19] Goldenbaum F, Morjean M, Galin J, Liénard E, Lott B, Périer Y, et al. 1999 Fission Time Evolution with Excitation Energy from a Crystal Blocking Experiment *Phys. Rev. Lett.* **82** 5012-5

[20] Barrué F, Basnary S, Chbihi A, Chevallier M, Cohen C, Dauvergne D, et al 2002 Nuclear fission time measurements as a function of excitation energy: A crystal blocking experiment. *Nucl. Instrum Meth.* B **193** 852-9

[21] Barrué F. 2002 Mesure de temps de fission par la méthode du blocage cristallin, Université Claude Bernard Lyon 1 (https://tel.archives-ouvertes.fr/file/index/docid/45013/filename/tel-00001888.pdf)

[22] Morjean M, Jacquet D, Charvet J L, L'Hoir A, Laget M, Parlog M, et al 2008 Fission Time Measurements: A New Probe into Superheavy Element Stability *Phys. Rev. Lett.* **101** 072701

[23] Grüner F, Assmann W, Bell F, Schubert M, Andersen JU, Karamian S, et al. 2003 Transverse cooling and heating in ion channeling *Phys. Rev.* B **68** 174104

[24] Stolterfoht N, Bremer J-H, Hoffmann V, Hellhammer R, Fink D, Petrov A, et al. 2002 Transmission of 3 keV Ne^{7+} ions through nanocapillaries etched in polymer foils: evidence for capillary guiding *Phys. Rev. Lett.* **88** 133201

[25] Cassimi A, Ikeda T, Maunoury L, Zhou C L, Guillous S, Mery A, et al. 2012 Dynamics of charge evolution in glass capillaries for 230-keV Xe^{23+} ions *Phys. Rev. A* **86** 062902

[26] Ray C, Bräuning-Demian A, Bräuning H, Chevallier M, Cohen C, Dauvergne D, et al. 2011 Measurements of high energy loss rates of fast highly charged U ions channeled in thin silicon crystals *Phys. Rev.* B **84** 024119

Journal of Physics: Conference Series **629** (2015) 012010 doi:10.1088/1742-6596/629/1/012010

[27] Toulemonde M, Balanzat E, Bouffard S, Grob J J, Hage-Ali M, Stoquert J P 1990 Damage induced by high electronic stopping power in SiO_2 quartz *Nucl. Instrum Meth.* B **46** 64-8
[28] Hijazi H, Rothard H, Boduch P, Alzaher I, Langlinay T, Cassimi A, et al. 2012 Electronic sputtering of LiF by Krypton (10 MeV/u): size dependent energy distributions of $Li^+(LiF)_n$ clusters *Eur. Phys. J.* **66** 305